LIGHT, VISIBLE AND INVISIBLE

AND ITS MEDICAL APPLICATIONS

LIGHT, VISIBLE AND INVISIBLE

AND ITS MEDICAL APPLICATIONS

Angela Newing

Gloucestershire Medical Physics Service, UK

ICP

Imperial College Press

Published by

Imperial College Press
57 Shelton Street
Covent Garden
London WC2H 9HE

Distributed by

World Scientific Publishing Co. Pte. Ltd.
P O Box 128, Farrer Road, Singapore 912805
USA office: Suite 1B, 1060 Main Street, River Edge, NJ 07661
UK office: 57 Shelton Street, Covent Garden, London WC2H 9HE

British Library Cataloguing-in-Publication Data
A catalogue record for this book is available from the British Library.

LIGHT, VISIBLE AND INVISIBLE, AND ITS MEDICAL APPLICATIONS

ISBN   1-86094-164-8

Printed in Singapore.

PREFACE

There have been strong links between the pure sciences and medicine since the Scientific Revolution. This began in the late 16th century when the medieval notions of the earth being the centre of the universe and the human body somehow working by divine magic were gradually overcome.

In 1603, Heironymous Fabricius, a scientist in Padua, Italy, published a description of the valves in veins. Shortly afterwards, William Harvey, physician to James I, discovered the circulation of the blood and that the heart was merely a sophisticated pump. Luigi Galvani and Alessandro Volta, Italian physicists, famous for their construction of the first batteries in 1800, recognised the existence of electric currents and found that minute electric currents controlled nervous impulses and the movement of muscles. A century later, Sir Humphry Davy investigated the therapeutic effect of various gases, and demonstrated the anaesthetic effect of nitrous oxide (laughing gas). His assistant Michael Faraday, who subsequently replaced him as director of the Royal Institution and whose greatest achievements were in electromagnetism, worked on a number of applications of physics to medical science such as the development of optical glass. By removing impurities, he produced lenses which transmitted a high proportion of the visible spectrum. This introduced major developments in microscopy and allowed opticians to manufacture sophisticated aids to vision. The 19th century also saw the invention of such diagnostic instruments as the stethoscope (1818), the laryngoscope and the thermometer by physicists.

In the 1880s, various muscle and nerve disorders were treated by connecting patients via tin electrodes into electrical circuits. This was said to give good results in some cases. A few years later patients were exposed to magnetic fields by being made to lie on an insulated couch within a solenoid of stout copper wire. The apparatus was called a cage of conduction and it was considered to be an effective treatment for neuralgia, angina, gout, lumbago, sciatica, and rheumatism, as well as for various non-specific problems such as insomnia. S. Kuznitzky in Germany and P. Ishewsky in Russia noted skin changes and increases in blood pressure while using variable magnetic fields on patients in these cages, but no measurements of field values or treatment durations seem to have survived. One suspects that none were made.

The events which introduced physicists in numbers, and eventually led to the establishment of departments of Medical Physics in most large hospitals in the developed world, began just over a century ago. In 1895, Wilhelm Röntgen, professor of physics at Würzburg in Germany, produced X-rays, and a few months later Henri Becquerel, professor of physics in Paris, discovered radioactivity. The significance for medical science of both of these important happenings was quickly realised. Physicists were needed to construct and install X-ray apparatus, to try to measure the radiations involved, and, after the damaging effects of ionising radiation had been recognised, to design protective barriers.

Nobel Prizes were established in 1901 in accordance with the will of Alfred Nobel. He was a Swede who had made an immense fortune from making explosives and exploiting the oil fields between the Black and Caspian Seas. When he died in 1896, he left a very large sum to establish a Trust to distribute annual prizes in the fields of physics, chemistry and physiology or medicine. He also established prizes for 'idealism' and 'fraternity between nations'. Because his relatives contested the will, Nobel Prizes were not awarded until 1901. The first Nobel Prize for physics was awarded to Röntgen in 1901, and the 1903 prize went jointly to Becquerel and Marie and Pierre Curie for their discoveries in the field of radioactivity. Marie Curie was also awarded the 1911 Nobel Prize for chemistry.

More recently, Godfrey, (later Sir Godfrey) Hounsfield, a physicist working in industry, developed Computed Axial Tomography. This is now called CT Scanning. He cautiously suggested, when introducing this development to the medical science community at the 1972 British Institute of Radiology Congress, that "It is possible that this technique may open up a new chapter in X-ray diagnosis". This was probably the greatest step forward in medical physics since Röntgen's original discovery, and won for Hounsfield and

Allen Cormack, a South African physicist with similar ideas, the 1979 Nobel Prize.

The chapters which follow take the reader through the various discoveries and developments of radiation physics applied to medicine and bring each of them up to date, explaining the theory on the way. Many of the developers and discoverers were honoured for their work, and there was a liberal distribution of Nobel Prizes among them.

Modern medical physics encompasses non-ionising radiations too. Lasers, ultraviolet radiation, ultrasound, magnetic fields and so on are dealt with also using the same historical approach. The bioengineering areas are not covered, neither is medical computing, which is often the province of Medical Physics, but the reader is referred to several introductory books in these areas in the bibliography. A knowledge of physics up to GCE 'A' level is assumed, some anatomical or biological knowledge would also be helpful to the reader, but it is hoped that the explanations are fairly straightforward.

Historical details, which are not strictly required in order to understand the subject matter but which led up to the development of the subject, appear in a different typeface. By skipping these passages, the reader will still gain the appropriate knowledge of the subject of medical radiation physics, but will miss a lot of the interest.

"Light, Visible and Invisible" was the title chosen by Professor Silvanus P. Thompson, F.R.S. for his series of Christmas Lectures delivered at the Royal Institution, London, in 1896. Enormous progress has been made during the intervening century, but this still seems an excellent title for this book.

Angela Newing
April 1999

ACKNOWLEDGEMENTS

I am grateful to the Department of Physics of Imperial College, London for having given me the opportunity to convert the substance of a guest lecture, which I delivered in 1997, into a book. In the process I have been able to increase the text by a couple of orders of magnitude and, at the same time, bring the material up to date.

I have been greatly helped and encouraged by a large number of people. This list is not complete, but I acknowledge particularly the assistance given by the following people:

Dr Tony Bennett, curator of the Monica Britton museum and exhibition hall at Frenchay Hospital, Bristol, rekindled my interest in the history of medicine and has supported my endeavours in the history of medical physics with enthusiasm and practical advice. The E. M. I. Corporation provided a picture of Sir Godfrey Hounsfield's first X-ray CT experimental apparatus. The Department of Radiology at Gloucestershire Royal Hospital loaned slides of modern ultrasound and MRI scans. Ms Fiona Leppard, from Health Promotion Gloucestershire has managed to convert my scrappy line drawings into effective book illustrations. Professor John Mallard, from Aberdeen, kindly loaned photographs of his early experiments in nuclear magnetic resonance. Dr Dick Mould generously allowed me to use pictures from his extensive collection of historical illustrations which have appeared in his books. Dr Bill Vennart, from the University of Exeter, loaned some lecture illustrations. The Wellcome Foundation, through its chief archivist Mrs Julia Sheppard, provided photographs of Wilhelm Conrad Röntgen's early X-ray experiments.

Finally, I am grateful to the members of my Gloucestershire Medical Physics Service who I badgered with queries and who gave useful information and advice.

CONTENTS

CONTENTS

Chapter 1

X-RAYS AND RADIOACTIVITY

The Discovery of X-rays

On 8th November 1895, Professor Wilhelm Röntgen was carrying out experiments in his laboratory in the University of Würzburg. He was passing a high voltage current from an induction coil through an evacuated Crookes tube covered in black paper, when he noticed a glow on a fluorescent screen which was on the bench nearby. What distinguishes a good physicist from a casual observer is that the physicist does not dismiss such a phenomenon as a trick of the light but seeks to explain it by further investigation. The history of science is liberally peppered with discoveries which were made by this method. Röntgen noticed that the glow was greatest nearest to the anode of the tube so he concluded that some radiation which was invisible to the eye but which could penetrate both glass and paper, was being given off by the anode. He called this radiation X-rays because he had no idea at that time of its exact nature. He cautiously remarked to a colleague, Dr Boveri, "I have discovered something interesting, but I do not know whether or not my observations are correct." It was not until 1912 that Max von Laue, through his work in crystallography, proved conclusively that X-rays were part of the electromagnetic spectrum with wavelengths around one thousandth of those of visible light.

The necessary properties involved in the production of X-rays are high vacuum, a high voltage current and a discharge tube with suitable geometry. These were available to Röntgen because of the historical developments described below.

Vacuum Technology and Elementary Electricity

The earliest pumps had been devised in the first few years of the 17th century to remove water from mine workings, but the first attempts to create an "empty space" or vacuum by removing gas from a gas tight container were made independently by Evangelista Torricelli in Italy and Otto von Guericke in Germany.

Torricelli was one of Galileo's research assistants in Florence who thought that the atmosphere ought to exert some pressure. Towards the end of the year 1643, he inverted a glass tube full of mercury with its open end immersed in another vessel full of mercury and found that the mercury column sank to about 75 cms above the reservoir leaving a vacuum above it. He had thus invented the barometer, and, in his honour the unit of measurement of vacuum was named the torr. Five years later, Otto von Guericke in Magdeburg, Germany, invented a mechanical air pump which could reduce pressure to about 10^{-3} atmospheres. He was a keen astronomer who had concluded that the stars and planets must be moving in empty space since any air resistance would slow them down and there was no evidence of this happening. He wanted to form a vacuum in order to study celestial conditions, but had no knowledge of Torricelli's work. His apparatus consisted of a cylinder with a piston and a valve which he subsequently perfected for gravitational and atmospheric pressure studies.

In the 1660s, Robert Hooke made a pump, based on von Guericke's principle, which was used by his colleague, Robert Boyle, to demonstrate that electric and magnetic attraction could occur in a vacuum.

Jean Picard, a French priest and amateur astronomer, came upon an interesting discovery in 1678. While carrying a mercury barometer up some steps one dark evening, he noticed a glow within the vacuum above the mercury and found that it was possible to produce this effect whenever the tube was shaken. He wrote notes on this but he took the experiment no further.

Twenty years later, Jakob Bernoulli, a German professor of mathematics, read of Picard's work and managed to produce considerable amounts of light by agitating tubes of mercury with and without a vacuum. He thought that he might have discovered a light source better than the existing candles.

Bernoulli's work came to the notice of Francis Hauksbee, an instrument maker and demonstrator at the London Royal Society in the late 17th century. He was able to show that light production in a vacuum was caused by friction between the mercury and the glass walls of the vessel.

Moreover, any substance which produced electricity by being rubbed was also capable of producing light in a vacuum from friction.

Hauksbee discovered electrostatic induction by demonstrating the transfer of a glow from an excited vacuum tube to an inactive tube nearby. He also found that his finger tips glowed when held close to an excited tube, and he developed apparatus which produced large electrical sparks. This was the first work connecting electricity with a vacuum and thus a significant development. In the years that followed, public demonstrations of these new phenomena attracted large audiences and popularised physics.

Early in the 18th century, Stephen Gray in London discovered that electric currents could flow along conductors and be stopped by insulators. Soon afterwards, Charles-François de Cisternay Du Fay, working in Paris, experimented with electric current. While Gray and his assistant Granville Wheeler transmitted electricity over more than 10 metres using fishing line, Du Fay discovered that electricity was of two different types which attracted each other but repelled themselves.

Benjamin Franklin, an American living in Europe in the 1740s, conducted his famous experiments during thunderstorms and was able to prove that lightning was an electrical discharge rather than fire. It was Franklin who suggested that the flow of electricity was a stream of very small particles. By 1745, static electricity was being stored in Leyden jars. The Leyden jar was a glass vessel with tin foil pasted both inside and out to about two thirds of its height. The lid was an insulator pierced by a metal rod which was in contact with the inner foil, and this rod transmitted electricity to the foil while the outer foil was earthed. A number of Leyden jars could be connected in series to form a battery. Large shocks could be generated from stored electricity capable of killing birds and animals. Some patients suffering from paralysis were treated with electric shocks, but without success and, as far as is known, without fatal results!

The Production High Voltage

Hans Christian Oersted, working in Denmark, was the first to discover a link between electricity and magnetism. In 1820 he noted that a magnetic compass needle was deflected when placed near a conductor along which electric current was flowing. André Marie Ampère, in Paris, found that conductors carrying current either attracted or repelled one another depending upon the direction of current flow. Michael Faraday repeated Oersted's and Ampère's work and studied current flow both in solids and in liquids, which he called electrolytes. He was able to conclude that negatively charged particles moved towards a positive terminal while positively charged ones moved to a negative terminal. He coined the words "anode", "cathode" and "ion", and in 1838 he was regularly able to demonstrate discharges in partially evacuated glass tubes.

Faraday was the first to describe electromagnetic induction in the 1820s having found that, if a coil of wire had a magnetic field passing through it, and this magnetic field changed, then a voltage was induced in the coil. The voltage only existed while the magnetic field was altering and the more rapidly it altered the greater the induced voltage. The voltage was also increased by having a larger coil with more turns of wire.

A changing magnetic field could be produced by moving a permanent magnet to and fro in the coil, although it was not possible to achieve very high voltages by this method.

Electromagnetism is the reverse effect of electromagnetic induction. Instead of the production of electricity by changing the magnetic field, electromagnetism is the production of a magnetic field by electricity. A current passing through a coil of wire produces a magnetic field directed along the axis of the coil, and a changing magnetic field can be created easily by switching the current on and off. This is the principle behind the induction coil which was the main source of high voltages until well into the 1920s. The design of the induction coil evolved over several years but is usually attributed to Heinrich Daniel Ruhmkorff who, in spite of a German name, was a mechanic in Paris.

Precursors of X-rays

As early as 1785, William Morgan in London was experimenting with electrical discharges in a vacuum by boiling the mercury in a Torricelli barometer tube. He noted that, with increasing boiling time, the light produced was first violet then purple, red, green and yellow, and that, finally, it became invisible. After Röntgen's "discovery", it became clear that Morgan had, unknowingly, produced X-rays.

Later, Faraday's work on discharges encouraged others. By the 1850s, vacuum technology had progressed as had glass blowing technique.

A skilled German glassblower, Heinrich Geissler, was an expert in the manufacture of glass vessels in a variety of complicated shapes required by physicists. In 1838 Geissler developed a technique for sealing platinum wires into such tubes and, using a sophisticated vacuum pump invented by Hermann Sprengel, he was able to reduce the air pressure to the order of 10^{-4} atmospheres. Sprengel's first pump is shown in Fig. 1.1. Subsequent models employed 3, 4, or 5 'fall tubes' of narrow bore which rapidly filled with mercury and produced a good vacuum in a short time.

Passing electric currents through Geissler tubes produced a luminous glow whose colour depended upon the type of glass used and the residual gas within it. Once the coloured glow was established, stratification occurred in the discharge and sometimes there was also a dark space near the cathode.

Cathode rays were first demonstrated by Julius Plücker and Johann Wilhelm Hittorf in Bonn in 1859. They used an "L" shaped tube like that shown in Fig. 1.2. In this tube the discharge was convincingly shown to be coming from the cathode, as it occurred in the arm of the tube containing the negative electrode. They showed that cathode rays moved in straight lines and that they generated heat and phosphorescence where they hit the glass of the tube.

Plücker and Hittorf's work was further developed by Sir William Crookes in England, who designed many different vacuum tubes to demonstrate different physical properties of cathode rays. He was a high profile scientist and a brilliant writer and demonstrator, and vacuum discharge tubes thus became known as Crookes tubes. One of Crookes' experiments used a mica cross attached to the anode in a tube with which he demonstrated a clear shadow of the cross on the glass beyond the anode, incidentally demonstrating that cathode rays travelled in straight lines. It was Crookes who caused a paddle wheel of mica, placed between the cathode and the anode, to be rotated by cathode rays; he also focused the rays from a concave cathode to generate heat in a sample of metal.

At the 1879 meeting of the British Association for the Advancement of Science held in Sheffield, Crookes described his researches on the dark space around the cathode, and postulated that atomic charged particles were passing through the vacuum. Meanwhile, German scientists, led by Hertz in Bonn, said that they were "disturbances in the ether". It was not until Sir Joseph John Thomson's discovery of the electron in 1897 that it was confirmed that cathode rays were electrons.

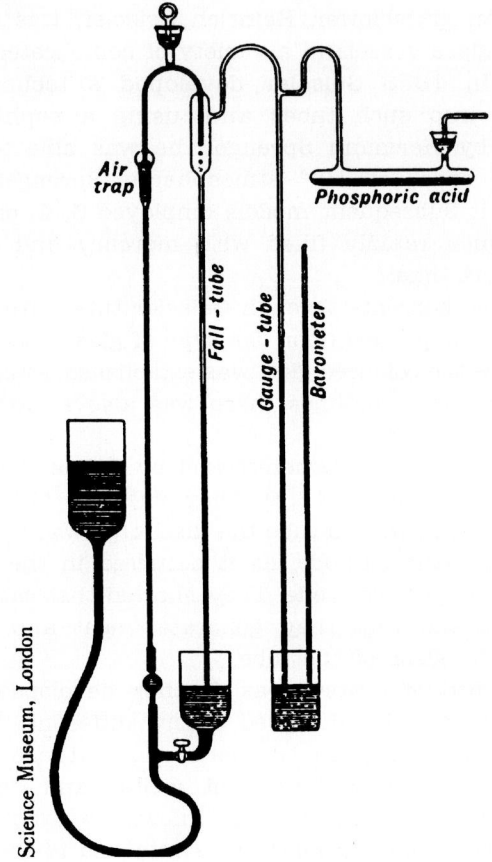

Fig. 1.1

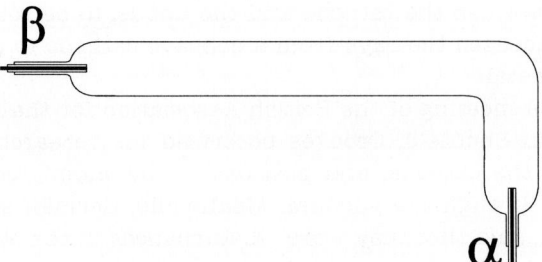

Fig. 1.2

Crookes probably produced X-rays during his experiments because he complained about the fact that a sealed box of photographic plates on his bench had mysteriously been blackened on one occasion.

Developments also continued in Germany. In 1894, Professor Lenard, an assistant of Hertz in Bonn, managed to detect cathode rays outside the envelope of a special discharge tube he had constructed. This tube had a thin aluminium window incorporated into the glass. Lenard's tubes were involved in addition to Crookes' tubes in Röntgen's experimental work of 1895.

Röntgen's Experimental Work

Having realised that his X-rays went through both glass and paper, Röntgen set out to find what else they would penetrate. He tried a wooden plank and a thick book, and then various pieces of metal. Figure 1.3, shows a reproduction of one of these early radiographs. Figure 1.4, is an X-ray of his shotgun.

During the month of November 1895, Röntgen continually repeated his findings to convince himself of the existence of X-rays. Unknowingly, he had protected himself from the harmful effects of radiation by the design of his experimental apparatus, having built a zinc walled cabinet with a small hole in one wall which he plugged with an aluminium disc. The X-rays came through the disc, and he added lead sheeting to this wall to prevent the passage of stray radiation. The cabinet was also his darkroom. These first X-rays were generated from the glass of the vacuum tube at the point where cathode rays struck the tube wall. Röntgen tried deflecting the cathode rays with a magnet and showed that the X-rays then came from the new site on the tube wall where the cathode rays impinged. The screen on which Röntgen first observed X-rays was coated with barium platinocyanide. Subsequently, he used glass photographic plates for his work.

Frau Röntgen worried about her husband's very long hours in the laboratory with little interest being shown in food or conversation. She eventually persuaded him to tell her about his discovery, and on 22nd December he demonstrated it to her by getting her to put her hand in the X-ray beam. The resulting photograph, reproduced in Fig. 1.5, was the first ever human radiograph. The exposure time for this picture was fifteen minutes.

Fig. 1.3

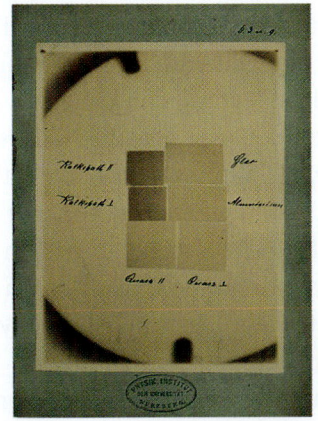

Fig. 1.4

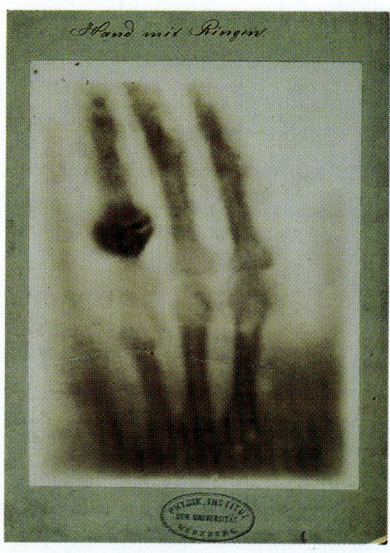

Fig. 1.5

Röntgen's first short paper on the phenomenon; "Uber Eine Neue Art von Strahlen" (on a new kind of ray) was published in the Proceedings of the Würzburg Physicalische Medicalische Gessellschaft on 28th December 1895 and distributed widely. Early in January 1896, Röntgen notified scientific colleagues in Germany and elsewhere, sending them translations of his paper together with copies of his radiographs.

On 23rd January 1896, Röntgen gave his first public lecture on X-rays in Würzburg University. Following the lecture, radiographs were passed around the audience and the X-ray apparatus was demonstrated using the hand of a volunteer. The resulting radiograph was received with loud applause. The meeting proposed that the new rays should be called Röntgen Rays and, although this name became popular in Germany, the name 'X-rays' had excited the general public elsewhere in the world, and became the name which was generally accepted world-wide.

Before long Röntgen's experiments were being repeated in Britain and America and the popular press had taken an interest. The 1896 Great Exhibition at Crystal Palace, London included an X-ray demonstration as did the Exposition of the Electric Light Association, organised in New York by Thomas Edison in the summer of 1896.

Photographic emulsions had been available since about 1825, but dry emulsions, employing gelatine coverings on the photosensitive chemicals had been produced in the 1860's. X-rays needed fast emulsions which soon became available from such suppliers as Britannia Ltd. (later Ilford Ltd.) in England and by Eastman Kodak in America. George Eastman invented a machine to mass-produce dry photographic plates which he patented in 1879.

The Development of the Photographic Process

Photosensitive chemicals were discovered early in the 18th century when a German chemist called Johann Schultz discovered that silver chloride and silver carbonate were discoloured considerably by light. The first permanent photographs were produced nearly a century later in 1826 by Niepce in France. Niepce's colleague, Daguerre, invented the first satisfactory photographic process in 1838 shortly before William Fox Talbot in England invented a different photographic technique to produce the sepia paper calotype.

In 1851, Frederick Scott-Archer produced wet collodion which had to be prepared into an emulsion immediately before exposure. Hill Norris soon discovered that a dry plate could be made by covering the collodion with gelatine. Improvements followed quickly, and photographic plates were in large scale production by the mid 1870s. By the time of Röntgen's discovery of X-rays, roll film, using a cellulose nitrate base was also available.

Early Medical X-ray Diagnosis

Medical radiology developed rapidly, partly because of major hostilities in the continent of Africa. The wars in the Sudan and elsewhere resulted in large numbers of casualties with severe gunshot wounds for which the treatment of choice was usually amputation of a limb or part of a limb, because it was normally impossible to locate the positions of bullets or shrapnel, and there was a strong likelihood of massive infection and gangrene when foreign bodies were left in wounds.

Transportable X-ray machines were first used in the Sudanese war in 1898. The apparatus employed a 10 inch induction coil and was operated by an army surgeon and an orderly. X-ray equipment was further developed for use in the Boer war of 1899–1902.

Meanwhile, the new science of medical radiology was developing in hospitals in Europe and America. In England there were X-ray departments in the capital at the London and King's College hospitals. At King's College, A. A. Campbell Swinton had repeated Röntgen's experiments in 1896 and S. Jackson set up the hospital department. There was much interest in Scotland, encouraged by Lord Kelvin, Professor of Natural Philosophy at the University of Glasgow, who had been one of the recipients of Röntgen's original communication in January 1896, and had been enthusiastic in passing on the new knowledge.

X-ray departments were often housed in the cellars or basement in damp accommodation which was poorly ventilated. This caused difficulties with the equipment, and the first part of every day was often spent in drying out the apparatus before a current could be passed through the X-ray tube.

The first attempts to show structures other than bones by the introduction of dense substances into the body, (later known as contrast media studies), took place in 1898 when mercury was injected into an artery of a cadaver to produce an arteriogram of the hand.

Fluoroscopy began in these early days, when X-rays from a continuously operating tube were directed at the patient, and a fluorescent screen of calcium tungstate was interposed between the patient and the operator to provide real time pictures of body structure.

Basic X-ray Physics

X-rays are produced when fast moving electrons are stopped suddenly by impact with a target. The kinetic energy of the electrons is converted largely into heat but also into X-rays which form part of the electromagnetic spectrum (Fig. 1.6). Photon energy is proportional to frequency, the constant of proportionality being Planck's constant (h) so that

$$E = hf.$$

Since frequency is inversely proportional to wavelength, so is photon energy. Hence,

$$E(keV) = 1.24 \, \lambda^{-1} (nanometres)$$

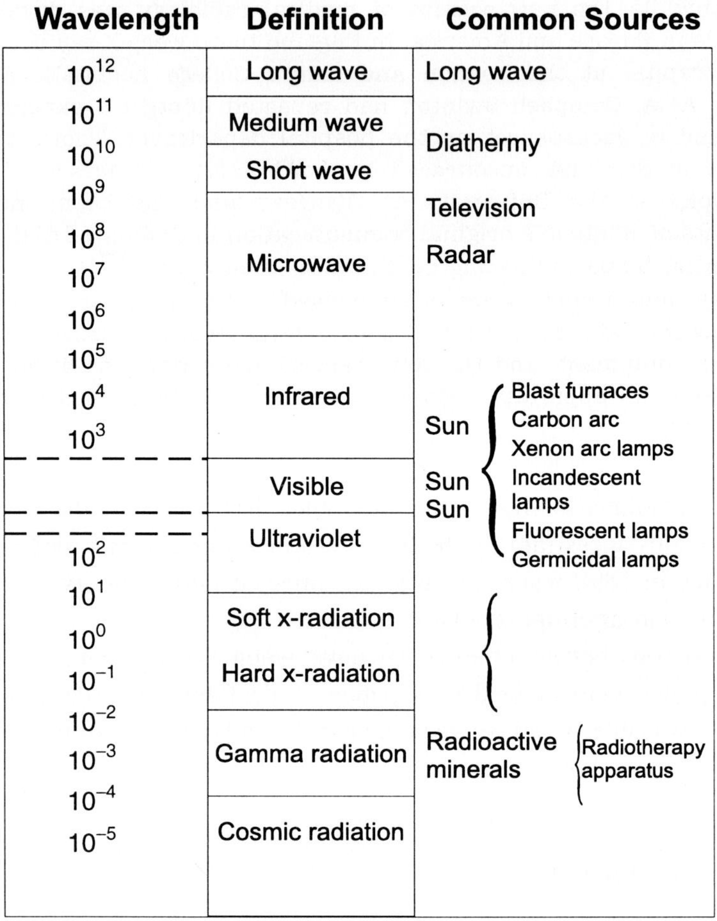

Wavelength	Definition	Common Sources
10^{12}	Long wave	Long wave
10^{11}	Medium wave	Diathermy
10^{10}	Short wave	
10^{9}	Microwave	Television
10^{8}		Radar
10^{7}		
10^{6}		
10^{5}	Infrared	
10^{4}		Sun — Blast furnaces, Carbon arc, Xenon arc lamps
10^{3}		
	Visible	Sun / Sun — Incandescent lamps
	Ultraviolet	Fluorescent lamps, Germicidal lamps
10^{2}		
10^{1}	Soft x-radiation	
10^{0}		
10^{-1}	Hard x-radiation	
10^{-2}		
10^{-3}	Gamma radiation	Radioactive minerals — Radiotherapy apparatus
10^{-4}		
10^{-5}	Cosmic radiation	

Fig. 1.6

For typical X-rays for which $E = 140\ keV$, $\lambda =$ about $0.1\ nm$.
(For orange visible light $\lambda = 600\ nm$, $E =$ approximately $2\ eV$)

The X-ray Tube

Electromagnetic radiation travels in straight lines called rays, and thus the dimensions of the X-ray beam are proportional to the distance the rays have

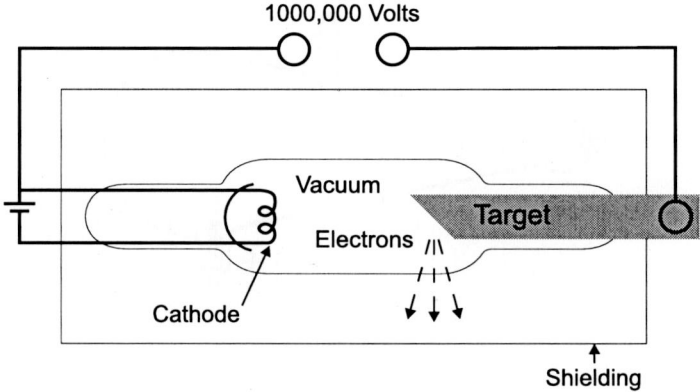

Fig. 1.7

travelled from their source. The area of the beam, and the intensity of the radiation, are inversely proportional to the square of the distance from the source.

Figure 1.7, shows the principle components of a modern X-ray tube. The cathode incorporates a filament which is normally a coil of tungsten, while the anode is a smooth flat metal target which is usually also of tungsten.

The Development of the X-ray Tube

The first X-rays, produced from a Crookes tube, were generated from the glass envelope when it was struck by cathode rays. The pear-shaped tube, (Fig. 1.8), produced X-rays from a large area of the curved surface giving unsharp images, also, because the glass became very hot, these tubes quickly deteriorated when used regularly.

Röntgen found that X-rays could be emitted from metals and that a suitable anode could be used for this purpose. The metal surface of the anode, which was normally made of platinum, became the target, and X-ray production became more efficient because of the higher atomic number. Experimenters then tilted the surface of the anode target so that the X-rays emerged at an angle and were not blocked by the cathode. In 1897, a concave cathode, like that used in cathode ray experiments by Crookes, focused the electron stream onto the anode and produced a sharper X-ray beam.

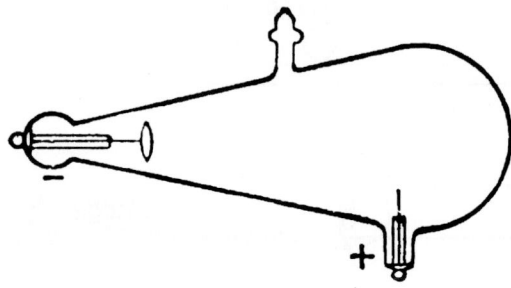

Fig. 1.8

The glass wall absorbed a proportion of the X-rays, and efficiency of X-ray production was improved with the developments shown in Fig. 1.9. Here, the tube thickness was reduced to 1 mm in the region where X-rays passed through it. The side chamber was added to increase the volume of the envelope in an attempt to prolong the tube life. Others successfully added an extra anode which had the effect of stabilising the X-ray output (Fig. 1.10). The residual gas in a tube was important. If there was very little gas, a higher voltage was needed to produce X-rays but the resulting radiation had greater powers of penetration.

Vacuum pumps based on von Guericke's method were improved and reintroduced (Fig. 1.11), but as tubes aged, the vacuum deteriorated. If the side chamber contained a small amount of potassium hydroxide, this absorbed any residual water vapour left in the tube when it was evacuated. As the gas pressure built up, the branch of the tube containing the potash could be heated to release some moisture which improved the vacuum. This was so successful that tubes were constructed where the heating of the potash was achieved automatically by a discharge spark which was triggered at the right moment. Figure 1.12, is a diagram of such a tube.

During 1897, an article in *Nature* contained a picture of 32 different types of tube (4). Early in the 20th century, tungsten was substituted for platinum as the target material. Its advantages of high atomic number, high melting point (3370°C as against 1773°C for platinum), high heat conductivity, and low vapour pressure at high temperatures, have stood the test of time and it is still the target material of choice.

An important development arose from the realisation that electrons emanating from the cathode activated purely by the electrical potential between anode and cathode, were insufficient to produce a stable

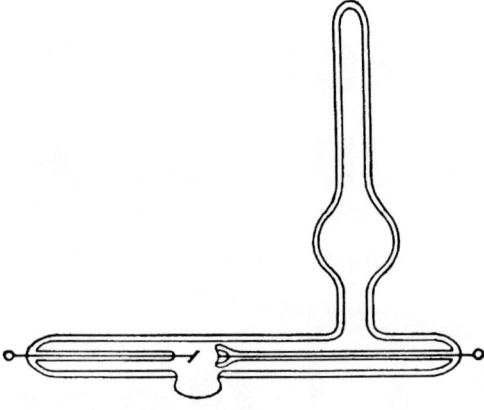

Fig. 1.9

Fig. 1.10

(by courtesy of the Monica Britton Museum, Frenchay Hospital, Bristol).

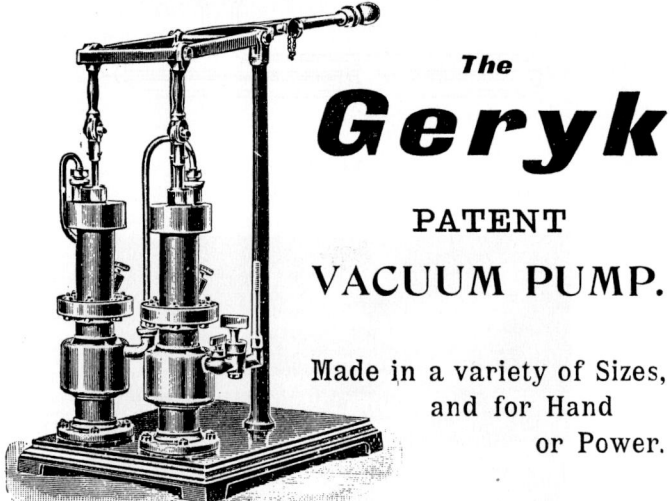
Fig. 1.11

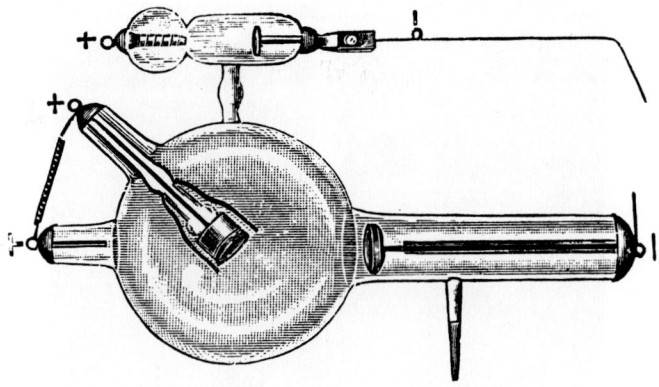

Fig. 1.12

X-ray beam in a near vacuum. Thomas Edison had invented the electric
light bulb with its incandescent filament some years before 1912,
when another American, Julius Lilienfield, decided to incorporate a
similar filament into an X-ray tube. This proved to be the breakthrough
required.

William Coolidge had worked hard to design a rugged and reliable
X-ray tube, and in 1913, he used Lilienfield's filament idea which
was immediately successful. A tungsten filament in the form of a flat
spiral was set within a molybdenum focusing tube to form the cathode.
This tube, known as the Coolidge tube, was capable of delivering
X-rays of uniformly good quality and large numbers of such tubes were
manufactured (Fig. 1.13). Filaments have been built into X-ray tubes
ever since.

More recent improvements, which are part of a continuous process,
have included reduction of focal spot size, the reduction of leakage
radiation by improvements in shielding, and electrical safety measures.
The first tubes to be fitted with rotating anodes were manufactured in
1929.

The Production of X-rays

As indicated in the foregoing section, modern X-ray tubes need two sources
of electrical energy.

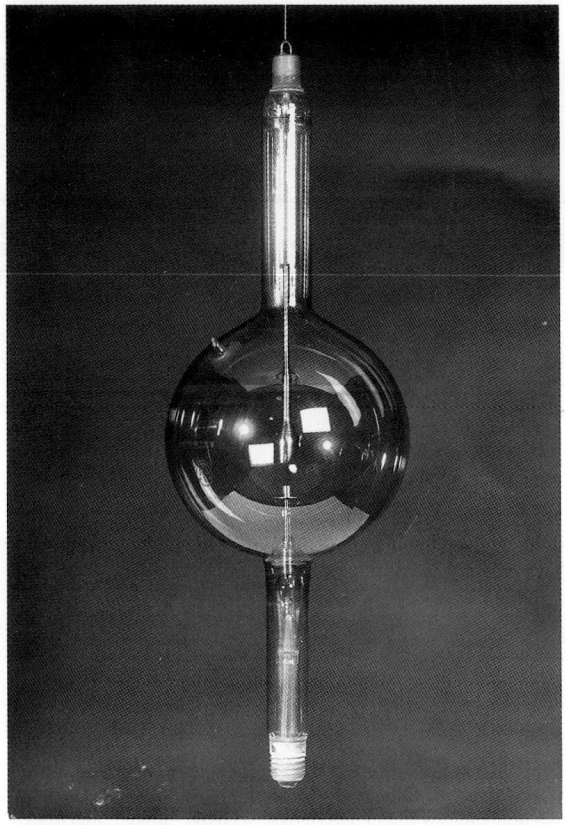

(by courtesy of the Monica Britton Museum, Frenchay Hospital, Bristol).

Fig. 1.13

1. The supply to the filament which is about 10 volts and 10 amps.
2. The supply across the tube between the anode and cathode, which is typically between 30 and 150 kv driving a current of electrons between 0.5 and 1000 mA.

The filament is heated to incandescence and, at such a high temperature (more than 2000°C), the agitation of atoms in the metal is sufficient to enable a small proportion of free electrons to leave the surface in spite of the attractive force of the positive ions. Once free, this "space charge" of electrons, as it is called, is repelled by the cathode and attracted by the anode which is the target. Because of the vacuum in the tube, the electrons arrive at the target with a velocity about half the speed of light and with kinetic energy equivalent to the voltage between the anode and cathode.

The X-ray spectrum emitted from the anode depends upon the kilovoltage of the tube and the target material, which is usually tungsten. A typical spectrum is shown in Fig. 1.14, which is a plot of intensity (the number of photons emitted) against photon energy.

One of three different processes occurs when electrons arrive at the target.

1. A bombarding electron may penetrate the inner k-shell of a target atom and approach the nucleus. Interaction with the nucleus causes an instantaneous change in the energy state of the electron, slowing it down and producing a single photon of X-rays known as

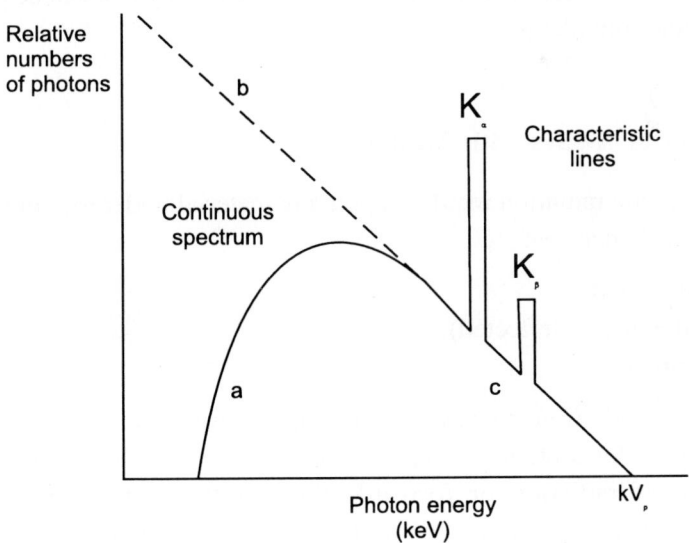

Fig. 1.14

Bremsstrahlung. Literally translated this is "braking radiation". This may have any energy up to the total kinetic energy of the electron. Bremsstrahlung produces the continuous part of the spectrum.

2. An electron may collide with an electron in the k-shell of a target atom, ejecting that electron if the energy of the incoming electron is greater than the binding energy of the shell. The 'hole' created is usually filled by an electron dropping from the l-shell with the emission of a single X-ray photon whose energy is equal to the difference in the binding energies of the two shells. This is called K_α radiation and produces a characteristic line on the spectrum. Occasionally, the hole may be filled with an electron from the m-shell with the emission of a characteristic line of higher energy representing the difference in binding energies between the two shells. This is called K_β radiation. There is also L radiation when a similar process occurs involving collision of the bombarding electron with an electron in the l-shell.

3. Very rarely, a bombarding electron may collide directly with the nucleus of a target atom and there occurs a total change of energy of the electron with the emission of a single X-ray photon equivalent to the peak kilovoltage. This is the largest photon energy that can be produced at this kilovoltage.

Interaction of X-rays with Matter

Electromagnetic radiation incident upon any material undergoes one of three interactions. It may be:

1. Transmitted.
2. Scattered (or deflected).
3. Absorbed.

Transmitted radiation passes through the matter unchanged. Scattered radiation may be scattered at any angle between travelling back towards the source of radiation or forwards through the matter, and absorbed radiation transfers all or part of its energy to the matter depending upon whether it is totally or partially absorbed. The proportions subject to the three interactions depend upon the energy of the incident beam and the content of the absorber.

The radiation emerging at the far side of the absorbing medium is called the attenuated beam, so attenuation is absorption plus scatter. The percentage attenuation for a narrow beam of X-rays or gamma rays is constant for constant energy and thickness of a given attenuator, and this means that attenuation follows an exponential law

$$\text{The intensity transmitted, } I_x = I_o \, e^{-\mu x}$$

where, I_o is the incident intensity, μ is the linear attenuation coefficient (the fraction of X-rays removed per unit thickness), and x is the thickness of the absorber.

The density of a substance, ρ, is defined as mass per unit volume, and is proportional to the number of atoms per unit volume. The probability of an interaction between an X-ray photon and a medium containing 'n' atoms per unit volume is also proportional to 'n'. Thus the total mass attenuation coefficient of X-rays in matter, (the fraction of X-rays removed from a beam of unit cross sectional area, in a medium of unit mass is represented by $\mu\rho^{-1}$.

Figure 1.15, Shows the three processes and demonstrates the exponential law of attenuation of X-rays in matter.

For a detailed discussion of the physics of the processes of absorption and scatter the reader is referred to Refs. 1 and 2, and for in-depth treatment of the subject, to Ref. 3. A brief description of the three important processes is as follows:

Photoelectric Absorption

An X-ray photon 'collides' with a bound atomic electron and transmits all its energy to that electron. This annihilates the X-ray photon and ejects the electron from its shell. The hole created is filled by an electron from elsewhere with the emission of a characteristic X-ray photon. The initial energy of the X-ray must be greater than the binding energy of the shell, and the characteristic X-ray photon, in the case of light materials

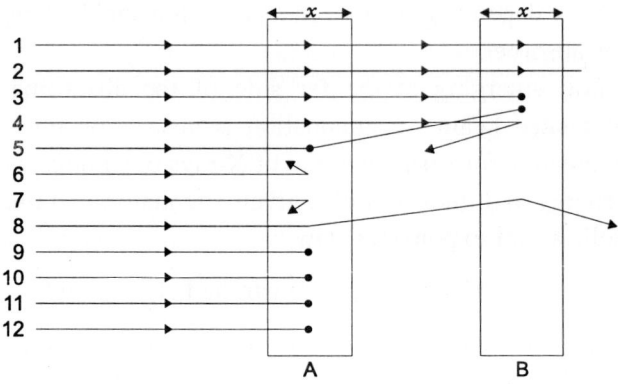

Fig. 1.15   Schematic diagram of the interaction of X-ray photons with matter.

like air and human tissue, is so 'soft' that it is immediately reabsorbed. Characteristic X-rays from high density materials are sufficiently energetic to pass through greater thicknesses and often escape from a patient being irradiated.

Compton Scatter

An X-ray photon interacts with a free electron giving up some of its energy in the process, but bouncing off in some other direction with reduced energy. The electron which was struck is projected either forwards or sideways while the scattered photon may bounce in any direction.

Pair Production

At high energies, an X-ray photon approaches the nucleus of an atom in the medium and gives up its energy to form an electron-positron pair. Energy is thus converted into mass.

The balance between the interactions means that low energy photons are preferentially absorbed by dense materials like bone, whereas, in water, fat and muscle, most of the radiation of this energy is transmitted. With increasing energy, attenuation in the materials making up the human body becomes almost uniform. This is shown in Fig. 1.16, from which it will be seen that X-rays below about 0.1 MeV are useful for diagnosis whilst higher energies are suitable for radiotherapy treatment.

The Discovery of Radioactivity

The discovery of radioactivity followed directly from the early X-ray work of Röntgen and other scientists who repeated the experiments. Fluorescence was observed in the glass of X-ray tubes and there was speculation that the X-rays themselves might be some sort of fluorescence or phosphorescence. Antoine Henri Becquerel was professor of physics at the Conservatiore National des Arts et Métiers and at the Musée in Paris when a colleague, Jules Henri Poincaré, suggested that it might be worthwhile to investigate the possibility of a form of X-rays being given off by phosphorescent materials.

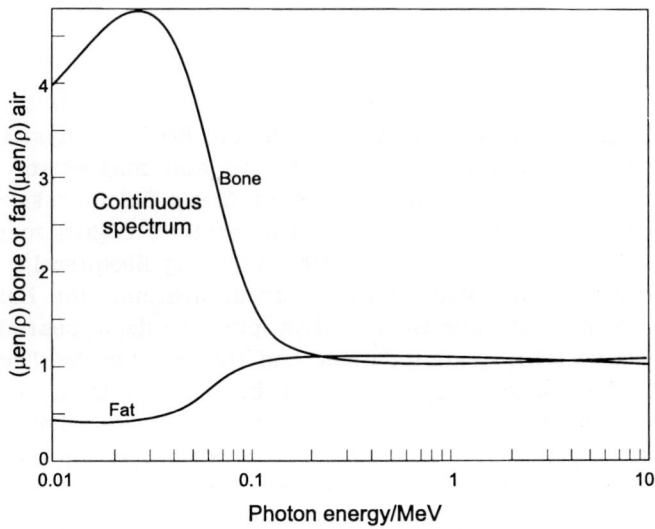

Fig. 1.16 Variation with energy of the ratios of the mass energy absorption coefficients of bone and fat to that of air.

Early in 1896, Becquerel undertook a series of experiments by exposing various fluorescent salts to sunlight, then wrapping them in black paper and trying to produce an image of a cross made of copper foil placed between the salt and a photographic plate. After a few attempts, he had succeeded in obtaining an image of the cross when using a fluorescent uranium salt. Like all good experimenters, he decided to repeat the experiment, but the day he chose, 26th February, was a typical cloudy February day, so he packed the salt and photographic plate away together in a dark cupboard.

It was still cloudy on 1st March, but Becquerel developed the plate anyway, and found an exceptionally strong image on it. Therefore, neither sunlight nor fluorescence was required to blacken the film and, by a happy chance, radioactivity had been discovered.

It is interesting to note that the British physicist, Silvanus P. Thompson, was also experimenting in this same area at exactly the same time. Early in 1896, he found a photographic action from uranium salts through an aluminium sheet which was thick enough to be impervious to X-rays. Coincidentally, his letter to the President of the Royal Society describing the phenomenon was dated 26th February 1896, but his work was not published until June that year.

Pierre Curie was a colleague of Becquerel's who taught at the Ecole de Physique et Chimie in Paris. In the 1870s and 80s his physics interests included infra red radiation, piezoelectricity and magnetism. In 1895, Pierre married one of his physics students, Marie Sklodowska who came from Warsaw in Poland. Marie received a master's degree in physics in 1893 and one in mathematics in 1894. Following Becquerel's discovery, Marie decided to investigate radiation from uranium for her doctoral thesis. The uranium ore she worked with, pitchblende, appeared to be far more radioactive than uranium itself and this led her to discover first polonium (named after her country of birth), and then radium. Her doctorate was awarded in 1903. Working with Pierre, they eventually isolated a whole gram of radium and studied its properties. This included the irradiation of various substances and of mice and guinea pigs.

Pierre Curie was killed in a road accident in Paris in 1906, after which Marie was appointed to the chair of physics which he had previously occupied. She continued her experimental work at the same time as bringing up two daughters. The Radiology Congress held in Brussels in

1910, established standards for radium in research and therapy, and this Congress defined the unit of radioactivity as the 'curie'.

Early Experiments with Radioactivity

During March 1896, Becquerel was able to demonstrate that the radiation from potassium uranyl sulphate could pass through thin aluminium. It was not reflected by a mirror or refracted by a prism but had the effect of discharging an electroscope.

In 1901, Becquerel and Pierre Curie published work which showed that one radioactive element could be transformed into another.

Following the discovery of radium, Marie Curie and Becquerel subjected the radiation from it to a magnetic field and demonstrated three categories of radiation:

1. Positively charged particles with dimensions similar to an atom. (alpha rays).
2. Negatively charged particles which were very light and were deflected very strongly in the opposite direction. (Beta rays or electrons).
3. Rays like X-rays which were not deflected. (Electromagnetic radiation).

We now know that radioactive decay can also involve positively charged electrons called positrons.

Radioactive Decay

Almost all atomic nuclei are stable and not subject to radioactive decay. Radionuclides are unstable nuclei which transform themselves spontaneously (decay) with the emission of some combination of alpha, beta, and gamma radiation until they reach a stable state. The disintegration causes the new nuclide to have a different atomic number, z, and sometimes involves a change of mass number also.

The law of radioactive decay states that the number of nuclei decaying per unit time (the rate of decay) is proportional to the number of such nuclei remaining. Thus, decay is exponential.

$$A_t = A_o \; e^{-\lambda t}$$

Where, A_o is the original activity, A_t is the activity at time t, λ is the decay constant for the radioisotope concerned.

A convenient unit, the half life, is the time for the original activity to decay by half.

$$T_{1/2} = 0.693 \; \lambda^{-1}$$

A few radioactive elements, mostly with very long half lives, exist in nature. An example is radium-226 with a half life of 1620 years. Its daughter product, the radioactive gas radon, is also in existence naturally although its half life is only 3.8 days.

Alpha, Beta and Gamma Rays

Alpha (α) rays consist of two protons and two neutrons bound together. This is equivalent to a helium nucleus. They are the least penetrating form of radioactive emission and are completely absorbed by a few centimetres of air or a very thin sheet of metal. However, because they travel at about half the speed of light and are brought to rest within such a short distance, their kinetic energy is lost with the production of many ions, and the effect is much more damaging biologically.

Beta (β) rays are electrons with a range of energies. They will pass through between one and two centimetres of human tissue, and their biological effect is twenty times less than that of alpha particles.

Gamma (γ) rays form part of the electromagnetic spectrum and generally have greater energies than X-rays. They are very penetrating and can pass right through the human body. They are usually absorbed in concrete or lead. The biological effect of X-rays and gamma rays is the same as for beta particles.

Other Work by the Curies

During the 1914–18 war, Marie Curie bought ambulances which she fitted out with portable X-ray equipment. Driving them herself, she became head of the French Red Cross. After the war she established 'curietherapy' — the treatment of cancer using radium, and helped to formulate safety standards for radiation workers. In 1921, the people of America gave her one gram of radium for use in curietherapy. (See Chap. 3).

Unfortunately, due to her long exposure to radiation, her own health deteriorated. She developed cataracts and skin lesions and suffered from anaemia. She died in 1934.

Marie's daughter, Irene, studied chemistry and physics in Paris and married another Parisian scientist, Frédèric Joliot. They experimented with bombarding light elements with emissions from radium and showed that some sort of radiation was emitted from the elements bombarded.

James Chadwick in England repeated these experiments using beryllium bombarded with alpha particles. He found that the particulate emission travelled at about one tenth of the velocity of light and thus was heavy. It was neither positively nor negatively charged, and he had discovered the neutron. Chadwick was awarded the 1935 Nobel Prize for physics, while the Joliot-Curies received the chemistry prize for their creation of phosphorus-30 from the bombardment of stable aluminium with alpha particles.

References

1. Farr R. F. and Allisy-Roberts P. J., *Physics for Medical Imaging* (W. B. Saunders, London, 1997) Chap. 1.
2. Wilks R., *Principles of Radiological Physics* (Churchill Livingstone, London, 1987, 2nd Ed.) Chap. 30.
3. Greening J. R., *Fundamentals of Radiation Dosimetry Medical Physics Handbook No. 15* (Adam Hilger, Bristol, 1985, 2nd Ed.) Chap. 2 etc.
4. *Tubes for the Production of Röntgen Rays. Nature.* **55**, 296–297, 1897.

Chapter 2

X-RAYS FOR DIAGNOSIS

Requirements for Diagnostic X-rays

X-rays arising from interactions in the inner parts of atoms in the target give a broad spectrum of radiation whose maximum energy depends upon the kilovoltage applied to the tube. Characteristic X-rays are of much less importance in diagnostic radiology than bremsstrahlung except in certain specialised techniques such as mammography. Higher kilovoltages give increased X-ray energy and greater penetration in body tissues, but with lower contrast (differential absorption) between different types of tissue.

Thus, for best effectiveness, diagnostic X-rays should have:

1. Penetrating power (kV) which can be varied over a wide range.
2. Intensity which is similarly variable.
3. Time durations which are very accurately controlled.
   (Very short times are sometimes necessary because voluntary or involuntary movements by the patient would blur the picture.)
4. A very small effective source (focal spot) to avoid geometric blurring which would arise with a large source.

Principles

As X-rays interact with, and pass through, the human body their attenuation varies depending upon the densities and atomic numbers of the intervening structures. As shown in Chap. 1, the linear attenuation coefficient is proportional to the number of atoms per unit volume, and this is one of the reasons why bone attenuates an X-ray beam more than soft tissue. The average densities of bone and soft tissue are 1.8 and 1.0 respectively.

What is considerably more important with X-rays of the relatively low energies used for diagnostic work, is differences in atomic number. In this energy range, materials of higher atomic number provide attenuation largely through photoelectric absorption, whereas Compton scatter is the dominant process for low atomic number materials. For bone ($z = 14$) photoelectric absorption is ten times greater than for soft tissue, (muscle $z = 7.5$, fat $z = 6.0$), because the photoelectric absorption coefficient at any particular energy is proportional to z^3.

The quality of early radiographs was poor for a large number of reasons:

1. The actual nature of X-rays was unknown at that time.
2. There was no realisation that low energy scattered rays caused 'fog' on films and fluorescent screens.
3. Fluorescent screens were not uniformly coated, and the materials used were not ideal.
4. There was no realisation of the necessity of matching photographic emulsions to the radiation quality.
5. Energy and intensity of X-rays were not controlled and were usually very variable.

There was much speculation as to the nature of X-rays. A quotation from a text-book of 1903 (1) is as follows:

"Wiechert estimates the velocity of an electron in its flight to be one third to one fifth that of light. In consequence of this enormous velocity the impact of an electron against a solid body sends an explosion-like elec-tric wave into space, just as a projectile at the moment of impact emits a sound wave. Wiechert thinks it not unlikely that Röntgen rays may be electro-dynamic wave movements, manifested as a series of short, rapidly-succeeding waves. The real carrier of these wave movements is the all pervading ether.

Other scientists regard Röntgen rays as cathode rays which have given of their charge at the tube wall or at the anti-cathode from whence they spring, thereby gaining in power of penetration. They are therefore cathode rays sifted, as it were, by the media through which they pass.

Assuming that Röntgen rays are themselves non-electrical, we can the more readily understand that on the one hand they tend to discharge an electrified body, and on the other hand, like all non-electric and non-magnetic bodies, they are not deflected by a magnet. Wehnelt's observation again, that the production of X-rays depends on no factor so much as the presence of rapidly suppressed discharges, that is to say, a disturbance of electrical equilibrium giving rise to powerful electric waves, induces others to assume that there is some connection between Röntgen rays and electrical waves of the most diminutive wavelength.

It is very likely that, just as white light is composed of various colours, Röntgen rays may be of various kinds, which differ from each other in their penetrative power as well as their physical and physiological effects."

The electromagnetic nature of X-rays was not finally established conclusively until 1912.

Scattered Radiation

When an X-ray photon is scattered, it loses some of its energy and changes direction. Scattered X-rays are of no use in diagnostic radiology as they do not contribute to the radiograph but increase the radiation dose to the patient. Some photons are scattered before they reach the patient but more are scattered by the patient's body.

Removal of Transmitted Scatter

A simple way of removing some of the scatter from a transmitted beam is to move the film some distance from the exit surface of the patient. As shown in Fig. 2.1, with distance, a proportion of scattered radiation emerging at wide angles from the patient will not reach the film, or will only reach parts of the film outside the main radiation beam. The effects of this are that film contrast is improved and the image is magnified. However, due to the increased distance between source and film, and the inverse square law, the kV or mAs must be increased, and some scatter will still reach the film and cause fogging.

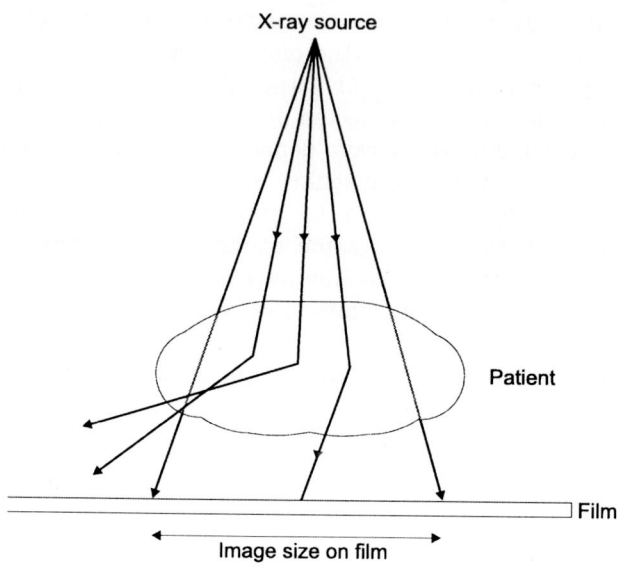

Fig. 2.1

In 1903, Otto Pasche in Switzerland designed a metal plate with a slit in it which travelled across the front of the X-ray plate during an exposure to remove scatter. At the same time another, synchronised slit moved similarly across the beam between the tube and the patient. Images were much improved but exposure times were greatly increased.

An improvement on this used a grid of heavy metal between the patient and the film. Figure 2.2(a) shows how transmitted radiation passes through gaps while some scattered radiation is absorbed in the metal mesh. Rays *a* and *b* are removed while *c*, which arrives at a greater angle, is transmitted. An increase in the width or height of the metal sections removes more scatter but at the expense of decreasing the beam intensity and increasing the weight of the grid. Scatter grids like this produced a pattern of grid lines on plates which sometimes obscured important structures.

The first moving grid was invented and patented by Gustav Bucky in Germany in 1913 and was developed jointly by Bucky and by Hollis Potter in America. The Potter-Bucky diaphragm or grid consisted of a system of parallel strips of metal which moved through an arc during the course of an exposure. This device became universally available in

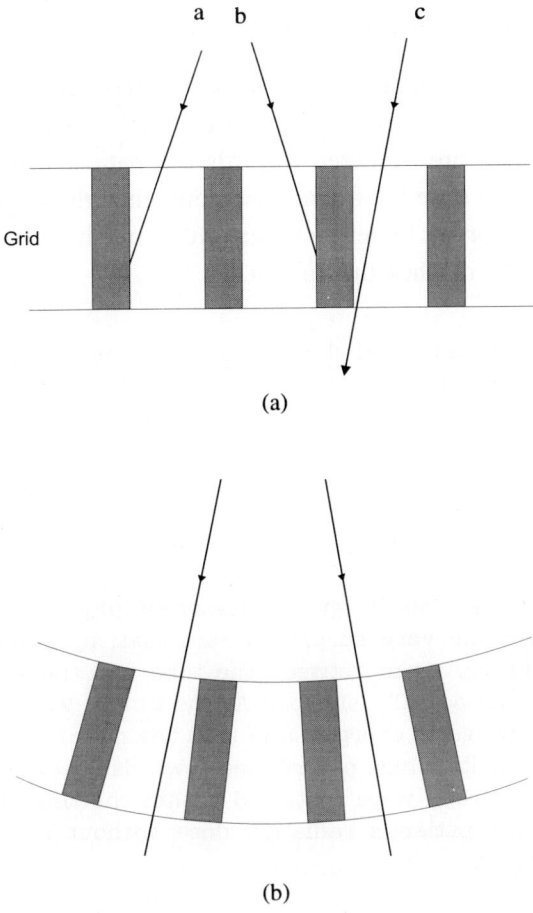

(a)

(b)

Fig. 2.2

the early 1920s and was the preferred method of removing scatter until the 1960s.

After 1960, fine line stationary grids with about 50 lines per cm became available and were found to be effective without producing grid lines on films. The focused fine line grid, as shown in Fig. 2.2(b), has been the method of choice for the removal of scatter since then.

Removal of Scatter from the Primary Beam

Some low energy photons are absorbed by the glass envelope of the tube (inherent filtration), others are scattered in the air between the tube and the patient. Those which are scattered towards the patient contribute towards the patient's radiation dose but are not energetic enough to cause blackening on the film. It is important therefore to remove as much low energy radiation as possible before it reaches the patient.

Early, non-focused, tubes radiated a wide band of energies in all directions. Fuzzy images were made worse by the scatter contribution from furniture, walls and other objects within the examination room. Focused tubes such as the Crookes 1897 concave cathode tube, and the angled anode tube, produced sharper images and, less than a year following the discovery of X-rays, various workers made cones, diaphragms and tube shielding to limit the size and direction of the useful beam.

In March 1896, William Magie, professor of physics at Princeton in America, described the very surprising fact that, by placing a piece of thin aluminium in the beam between the tube and the subject to be X-rayed, the image was actually sharper. At the time it was not known that X-rays covered a range of energies and that the metal was removing the low energy components which contributed towards scatter and fog. These low energy (soft) X-rays were scattered within the tissues, contributing substantially to the patient's radiation dose without having any effect upon the X-ray image.

Aluminium ($z = 13$) absorbs strongly at low energies by photoelectric absorption but there is very little Compton scatter. The small amount of characteristic radiation from the aluminium is insignificant and this is why aluminium became, and still is, the usual material used to harden a diagnostic beam. The thickness used is about 1.5 mm to 2.0 mm. Copper ($z = 29$) is occasionally used with higher kVp X-rays, but it emits characteristic radiation at 9 keV which must be absorbed by adding a 'backing filter' of thin aluminium on the side away from the tube.

Mammography

The breast does not attenuate an X-ray beam very much and mammography attempts to find small micro-calcifications within it. Ideally, X-rays of about 20 keV are needed, and the apparatus presently used employs one of two different arrangements. Most commonly, a tube operating at a constant potential of about 25 kv has a beryllium window and molybdenum target with the addition of a thin molybdenum filter. This combination transmits the characteristic radiation, at 17.9 and 19.5 keV, whilst removing almost all of the continuous spectrum. Slightly higher energy characteristic radiation, at 23 and 24 keV, can be provided by using a rhodium target and rhodium filter and is preferred for the examination of thicker breasts. This technique is not entirely satisfactory because very small calcifications are easily missed and some calcification can occur in benign tumours.

Breast tissue varies in consistency between patients. Younger women and those who are taking hormone replacement therapy have a high proportion of glandular tissue, but, with increasing age this is replaced by fatty tissue. The higher attenuation coefficient of glandular tissue makes it difficult to see calcifications in these patients.

A new technique involves high intensity synchrotron radiation. Synchrotron radiation is the name given to photons emitted when very high energy electrons or positrons are bent in magnetic fields. Radiation across the whole spectral range from infrared to X-rays up to several GeV can be produced.

X-ray diffraction using synchrotron radiation might provide better diagnosis in the future. Diffraction experiments on samples of collagen from normal breasts show close similarities between all samples. Malignant breast collagen may produce different diffraction patterns. It is thought that calcifications in malignant tumours might have a different crystal structure from those in benign lumps. However, the usefulness of this technique is currently limited to the examination of tumours which have been surgically removed. An added advantage of this emerging method of diagnosis would appear to be a very large reduction in radiation dose to the patient. However, both ultrasound and magnetic resonance imaging are very satisfactory modalities for *in vivo* breast examination, (see Chap. 6), and neither involves ionising radiation at all.

Fluorescent Screens

X-rays cause excitation of the atoms within certain materials called phosphors which causes them to emit light. Valence electrons are raised to a higher energy level where they are caught in energy traps before dropping back to their valence shells with the emission of light. When this happens instantaneously, the process is called fluorescence whereas, if there is a noticeable time interval, it is phosphorescence. Some materials require stimulation by heat before light can be emitted. This is called thermoluminescence and the process has no application in radiology although it is useful in radiation dosimetry. (See Chapter 3).

X-rays were discovered because of their effect upon a fluorescent screen of barium platinocyanide. This is not the only, nor necessarily the best, material for X-ray fluorescence however.

In 1896, Thomas Edison experimented with hundreds of different materials and found more than fifty which fluoresced when irradiated. The most efficient of those tested was calcium tungstate, giving bluish-white fluorescence, which he incorporated into a device called a fluoroscope. This was in the shape of a truncated pyramid with a narrow aperture for the eyes at the top and a screen of calcium tungstate at the wider end. The subject to be examined was positioned between the tube and the fluoroscope. It was important to exclude light from the fluoroscope because the brightness of the screen was too low to stimulate the cones in the retina of the viewer's eye (photopic vision). In near darkness it is the rods which provide vision (scotopic vision).

In England, barium platinocyanide was still used because the fluorescence, although not as intense, was bright green and more easily visible. Experiments with other fluorescent materials in the early part of the 20th century led to a better material, zinc sulphide, but this still provided insufficient brilliance for daylight viewing. Radiologists would wear red goggles for some ten to fifteen minutes before undertaking fluoroscopic investigations and the examination room was kept very dark.

Direct viewing fluoroscopy produced a number of 'radiation martyrs', workers who died as a result of their radiation exposure. The first

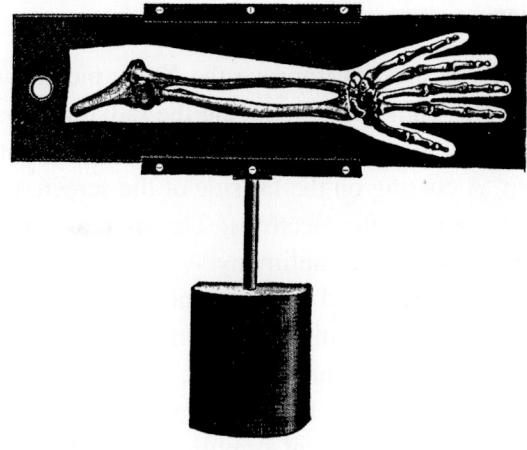

Fig. 2.3

was Clarence Dally, an assistant of Thomas Edison in New York. Dally was heavily involved in the X-ray exhibit at the 1896 Exposition of the Electric Light Association. He was responsible for switching on the apparatus each morning and it was customary for him to hold his hand between the tube and the screen while adjusting the circuitry to obtain the best picture. He also regularly demonstrated the use of X-rays elsewhere and was later involved in the examination of patients. In 1900, he developed ulcers on his hands which would not heal, then his hair fell out. He became very ill and died in 1904.

Following the realisation of the harmful effects of X-rays, a device called an osteoscope (Fig. 2.3), was constructed. This instrument consisted of an articulated skeleton of the hand, or hand and lower arm, mounted in contact with a fluorescent screen. The hand of the operator was shielded with a metallic semicircular plate when he held it in the beam while adjustments were made.

It became the practice of radiologists to protect themselves from the primary beam with lead glass in the form of a screen. This decreased the visibility of the image, leading to higher radiation doses to patients which would be totally unacceptable by modern standards.

These problems were overcome when image intensifiers were introduced.

Image Intensifiers

An image intensifier is used to improve the X-ray picture, making it easier to interpret, and, at the same time, reducing the radiation dose to the patient. The system works by firstly converting X-ray signals into light using a fluorescent screen. A coating on the far side of the screen, the photocathode, converts the light pattern into electrons. The most common materials for photocathodes are caesium and antimony.

The electrons are accelerated by a potential of about 30 kV towards a positive output screen via a focusing mechanism, the electrostatic lens. Another fluorescent material, zinc cadmium sulphide, is incorporated into the output screen to convert the image back to a visible one. A thin layer of aluminium lines the output phosphor to prevent stray light from entering the intensifier (Fig. 2.4). Although this may appear to be a complex process, it works very well and has enormous advantages over fluorescent screens alone.

The brightness of the output screen can be greatly increased electronically so that the observer's eye does not have to be dark adapted. The eye is better able to distinguish between levels of brightness when these levels are higher, so diagnostic ability is increased. The equipment operates in real time so a moving picture is produced. The image can be reduced or magnified and can be made to appear on more than one monitor screen. The operator can be located away from the primary beam. Modern image intensifier systems have

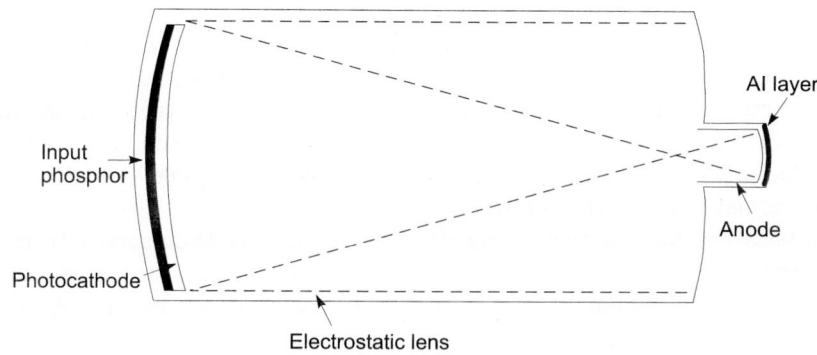

Fig. 2.4

the ability to "freeze" frames, allowing the radiologist to study a picture with-out continually irradiating the patient.

The Use of Films and Screens

Production of early radiographic images on film was based on the principle that photographic emulsion is sensitive to X-rays. A "fast" emulsion was required and the first radiographs were produced on glass plates coated with silver halide.

A description of the use of photographic plates in a text book of 1903 stated:

"Photographic plates are caused to fluoresce by the X-ray, but it has not yet been determined whether the photochemical action is a secondary one due to this fluorescence, or is produced directly by the X-ray." (2)

The usefulness of diagnostic X-rays depends upon their power to penetrate tissue, and this means that the beam emerging from the patient largely passes through any film positioned to capture the image. Only about 1% or 2% of the radiation is absorbed and the rest does not contribute to the picture.

From the first years of the 20th century, workers looked for ways of using this wasted energy at the same time as realising that photosensitive chemicals were more sensitive to visible light than to the X-ray part of the spectrum. This led to the introduction of intensifying screens in 1920.

An intensifying screen is designed to optimise absorption of X-rays and convert the X-ray energy into light which is absorbed by the film. Any phosphor will convert X-rays into light, but a phosphor with a high atomic number is most efficient.

An X-ray cassette consists, therefore, of a light-tight flat box fastened by spring clips. The film, which usually has emulsion coated on both sides, is sandwiched between a pair of intensifying screens and good contact between screens and film is maintained by pressure pads of low atomic number material. Single emulsion film is used when the small degree of unsharpness produced by double coated film would be unacceptable, for instance, in mammography.

The front of the cassette was normally made of aluminium to allow transmission of most of the incident beam while the back incorporates a sheet of thin lead to absorb residual radiation which might otherwise be subject to backscatter towards the patient reducing the quality of the image.

In the 1980s, carbon fibre ($z = 6$) began to be substituted for the aluminium on the entrance side, allowing more radiation to reach the film with a consequent reduction in patient dose. Carbon fibre is particularly useful at low kilovoltages and is therefore used most in mammography, angiography and orthopaedics.

The most frequently used screen material until about 1985 was calcium tungstate. At that time various rare earths were tried and lanthanum oxysulphide and oxybromide, and gadolinium oxysulphide were found to be more sensitive than calcium tungstate. The rare earths fluoresce at different wavelengths from tungsten making it necessary to use film with different characteristics in these cassettes. It is also important that safe lights of the correct colours are used in processing rooms.

The Heel Effect

A variation of intensity across the X-ray beam in the anode-cathode direction was first noticed following the general introduction of angled targets. This is called the heel effect and is illustrated in Fig. 2.5. X-rays are generated within the metal of the target and emerge at a variety of angles demonstrated by rays 1 and 2, in the diagram. The ray in position 1 traverses a greater thickness of the target than 2 and is thus attenuated more. It is also 'harder' because of the filtering effect of the target material.

The heel effect is not normally a problem in diagnostic radiology since small variations in image quality do not affect diagnosis. It can be used to advantage when non-uniform parts of the body are radiographed. For instance, for pelvic radiography, by positioning the cathode end of the tube towards the thicker, lumbo-sacral area and the anode towards the pubis, a

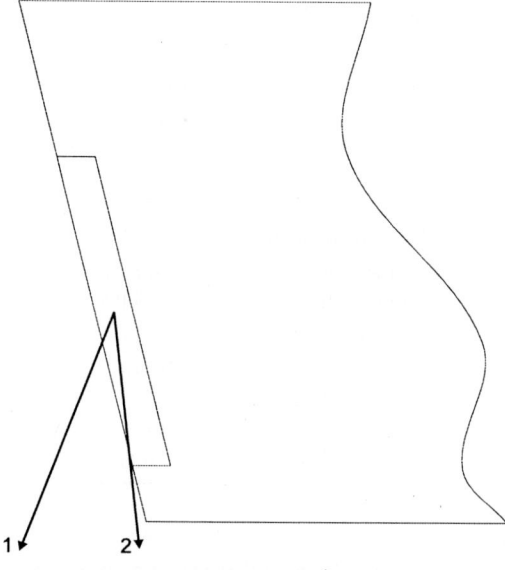

Fig. 2.5

more even image is obtained. The heel effect is particularly useful in mammography.

The Use of Contrast Media

There is very little difference between the attenuation coefficients of soft tissues, Table 2.1 shows some attenuation coefficients of human tissue for X-rays in the diagnostic energy range (2).

From this it can be seen that straight radiographs are of limited use in visualising structures other than bones. Contrast media are substances which are relatively opaque to X-rays and which can be put into the body to outline soft tissue structures.

As long ago as January 1896, attempts were made to outline veins and arteries, when the severed hand of a cadaver was injected with mercury and a good map of the arteries was seen. Later that year a solution of

Table 2.1

	Coefficient μ (cm^{-1})
Fat	0.185
Water	0.203
Cerebrospinal fluid	0.207
White matter	0.209
Grey matter	0.211
Muscle	0.212
Blood	0.213
Blood clot	0.221
Bone (average)	0.380
Cortical bone	0.629

chalk was used as was calcium sulphate on post mortem specimens. Radiography of blood vessels is called angiography.

The first human angiogram on a living volunteer was performed by Carlos Heuser in Argentina in 1919, the same year that Jacobaeus in Sweden produced the first human myelogram in which the spinal canal was outlined with a contrast agent. Lipiodol, a patented formulation with a high iodine content of about 0.3 grams per millilitre, was introduced in 1921 for bronchography and myelography, but it was subsequently found to have unacceptable side effects.

In 1923, Leonard Rowntree accidentally discovered how to radiograph the function of the renal tract. He had injected a patient suffering from syphilis with intravenous sodium iodide and saw the opacification of the kidneys and ureters on a subsequent radiograph.

In 1927, a Portuguese radiologist, Egon Moniz, was awarded the Nobel Prize for Medicine after producing the first cerebral arteriogram using sodium iodide. He later developed the use of thorium dioxide suspension for this technique.

Following the general introduction of contrast studies, some established radiologists felt that their use in diagnosis was a form of cheating. Clinicians who resorted to contrast techniques were looked down upon. As late as 1940, it was said that, if a neuro-radiologist had to resort to cerebral angiography, he did not know how to examine his patient properly. Fortunately, this view did not persist for long.

Contrast Materials

The most popular compounds for contrast studies contain high proportions of either barium or iodine. The qualities needed are that the element concerned must have a sufficiently high atomic number so that photoelectric absorption is maximised but the characteristic X-rays must be sufficiently energetic to be able to pass through the patient so that they perform like Compton scattered rays. The k-absorption edge needs to be just slightly below the peak of the X-ray spectrum transmitted through the patient. Since this peak is usually around 40 keV, barium ($z = 56$, $E_k = 37$ keV) and iodine ($z = 53$, $E_k = 33$ keV) are particularly suitable. The form of the compound for barium is usually an aqueous suspension of barium sulphate, and for iodine various liquid organic compounds containing high percentages of iodine. For studies of the gastrointestinal tract, air is also a useful contrast medium.

In addition to angiography, arteriography and the other studies mentioned above, cholecystography studies the gall bladder and bile duct, organs which are difficult to access except surgically, because they lie surrounded by the liver and the gut.

Cholecystography was first attempted in 1924 using a derivative of phenolphthalein. This caused enormous gastrointestinal upsets in patients, but it did outline the bilary tract well, and the side effects were tolerated until 1940 when pheniodol was introduced, followed in 1951, by iopanoic acid which contains 66% iodine and causes few side effects.

After injection, the solution is excreted by the liver. It pours down the bile duct and overflows into the gall bladder a few hours after injection. Radiographs taken at this point show the outline of the gall bladder. If the patient is then given a fatty meal, the bile runs int the intestine and the bile duct is outlined on X-ray. Gallstones may be opaque or non opaque and can be located in the gall bladder or one of the ducts.

For renal studies, the contrast medium is injected into a vein. It concentrates in the kidneys first from where it is excreted via the ureters to the bladder and urethra. Thus, by taking sequential radiographs, abnormalities of any of these organs may be diagnosed.

Tomography

The text so far has described the production of images straight through the patient. A single radiograph of this type is not particularly useful for some types of examination, for instance, for the location of radio-opaque foreign bodies. The only indication of the depth of such an object is its apparent magnification on the film.

The production of two radiographs at right angles to each other allows the position of an object to be located, but movement of the patient between the exposures introduces errors. However, this method is especially useful for the location of radioactive sources used in radiotherapy when the patient is anaesthetised or sedated, and a very accurate 3-D reconstruction of an implant is necessary for dose calculation.

Another method of location, which was available early in the 20th century, involved the production of a pair of radiographs in the same plane but with the tube moved through a known distance of a few centimetres to represent the view from each eye. These could be recombined stereoscopically to make the image appear to be three dimensional, and various devices were invented to reconstruct the geometry accurately as shown in Fig. 2.6.

X-ray slices through a patient are extremely useful. A linear method of obtaining slices was developed about twenty years after X-rays were first used. This technique uses a tube and film holder joined by a bar so that the pair rotate through an arc around a point. The body structures in the plane of the centre of rotation are imaged sharply on the film whilst everything above or below this plane is blurred. The principle is illustrated in Fig. 2.7.

Linear tomography was invented independently by about ten different workers, and patents applied for in at least five countries. In 1913, Karol Mayer in Poland whose main interest was in chest radiology, moved his tube back and forth to blur the ribs but show the heart. A year later, Baese an Italian engineer, used a similar method to locate bullets. More sophisticated apparatus was developed in France in 1921 by Bocage and by Portes and Chaussé. In the later 1920s, several different methods of moving the tube and film were tried including circular, spiral and cycloidal.

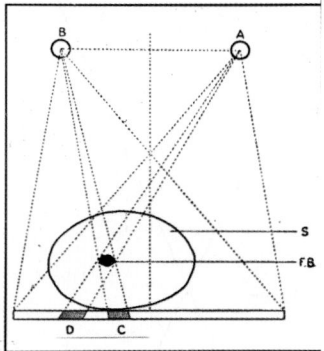

Fig. 2.6

Tomography was also called stratigraphy and planigraphy by early workers, but tomography became the accepted term when the technique became widely available in the mid 1930s.

In 1938, G. B. Watson and I. Valerian achieved transverse sections through the body, but further development was interrupted by World War 2.

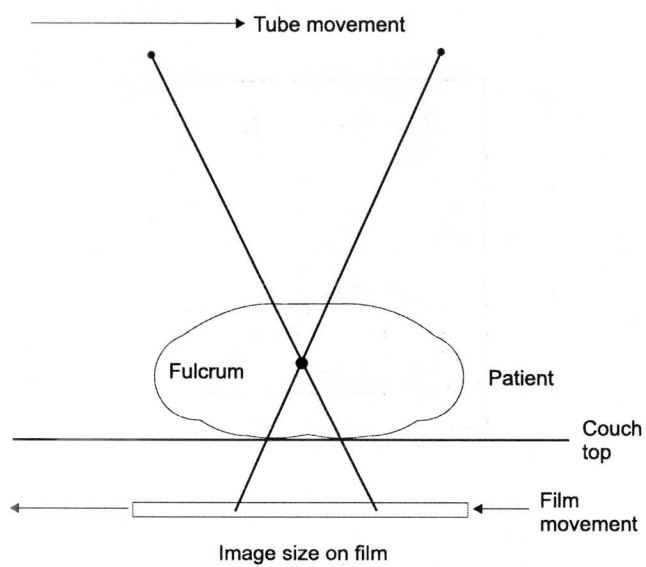

Tube movement

Fulcrum   Patient

Couch top

Image size on film

Film movement

Fig. 2.7

A variation, called pantomography, was invented in 1949 to show the teeth and the whole jawbone in focus. This involves rotating a film and tube horizontally around the head of a seated patient. It is of great value in dentistry.

Computed Tomography

Basic Concepts

In Chap. 1 the equation for transmitted intensity (I_x) was derived.

$$I_x = I_0 e^{-\mu x}$$

where x is the thickness of the absorber and m is the linear attenuation coefficient of the absorber. When the X-ray beam passes sequentially through different materials with thicknesses x and y, the transmitted

$$I_{(x+y)} = I_0 e^{(\mu x + \mu y)}$$

and, in general for many different material

$$I_t = I_0\, e^{-\Sigma \mu t}$$

Godfrey (later Sir Godfrey) Hounsfield was an engineer working for EMI Limited in London when, in 1968, he felt that conventional X-ray methods were using the available information only in a limited fashion. He thought that, by taking enough measurements and analysing them by computer, it should be possible to calculate the individual attenuation coefficients of all the body structures. It should also be possible to produce cross-sectional slices which would show the depth of objects and organs within the body. Additionally, information from consecutive thin slices should allow 3-dimensional reconstruction.

He experimented first by rotating a narrow beam from a gamma ray source opposed by a detector, around a section of human brain. He was able to set this up on a lathe bed and took measurements at intervals of 1° around the specimen. A map of the cross section of the brain was divided into a matrix of small volume elements called voxels, and by calculating which voxels each particular pencil beam traversed, it was possible to reconstruct a matrix showing the individual attenuation co-efficients of the voxels. The computer then produced a picture with a corresponding matrix of picture elements called pixels giving a detailed differentiation between the structures. Hounsfield was able to map the areas of grey and white matter separately on his first pictures, but it had taken several days to acquire the data.

By substituting a narrow X-ray beam as the radiation source, the time was reduced to nine hours, but he gradually perfected the technique so that it was feasible to build a prototype for use with patients. Figure 2.8, shows Hounsfield's prototype apparatus.

A few years earlier, in 1964, Allen Cormack, a South African radiotherapy physicist, had developed an interest in the effects of inhomogeneities with-in the body on radiotherapy dose distributions. He attempted to produce a cross sectional image from a series of angular projections of an X-ray beam at 7.5° intervals around the body. His pictures were of poor quality but, since his was the same basic idea as the one developed by Hounsfield, the pair of them shared the 1979 Nobel Prize for physics.

Hounsfield published his early results in 1972 and the first prototype EMI scanner was installed at Atkinson Morley Hospital, London in that

© EMI ARCHIVE
SCANNER (CT PROTOTYPE)

Fig. 2.8

year. The patient had to be scanned for about five minutes and further time was required to reconstruct the image. This machine employed a pencil beam of X-rays and was only useful for examinations of the brain. The second generation EMI scanner employed a 10° fan beam of X-rays opposite an array of detectors. This reduced the scanning time to about 20 seconds and improved the quality of the image.

A further improvement in 1975 used a fan beam of X-rays of about 25°. The fourth generation scanner in 1980, had a stationary array of over 1000 detectors surrounding the patient. All these developments increased the scanning speed, improved the quality of the images and, eventually, allowed cross sectional images of the thickest parts of the body to be obtained.

Reconstruction of CT Images

The method used to produce CT slices is called filtered back projection. It can be described as follows.

Figure 2.9(a) represents two radiographs of a radio-opaque cylinder taken at right angles to each other. On the two films, areas 'b' and 'e' will appear as dark bands whereas 'a', 'c', 'd', and 'f' will be clear.

If visible light is shone through the two films from behind, as in 2.9(b) a band of shading is projected from the central portions of each film. These cross in the region 'x'.

By increasing the number of projections, as in 2.9(c), the cross section of the cylinder can be reproduced accurately because the degree of blackening at the centre is far greater than that elsewhere.

Figure 2.10 shows how this principle works for two objects at positions 'x' and 'y' when four projections are added together.

The technique requires mathematical modification when fan beams are used, but computed reconstruction calculations are very fast.

A unit called CT number, (originally called Hounsfield number), has been developed for CT scanning which compares the linear attenuation coefficient of pixcels of tissue with water. Since the energy dependence of m_{water} and

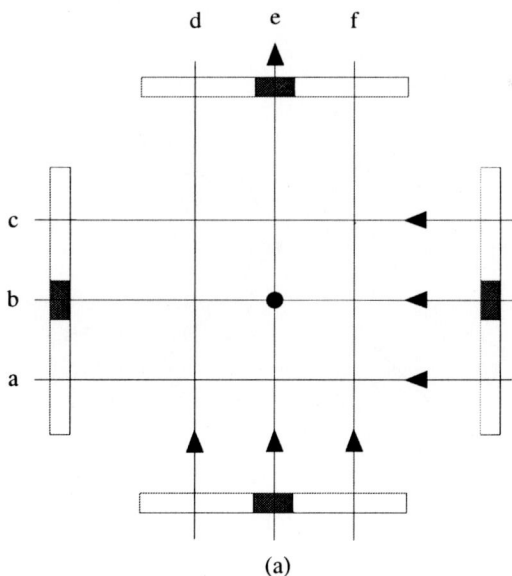

(a)

Fig. 2.9

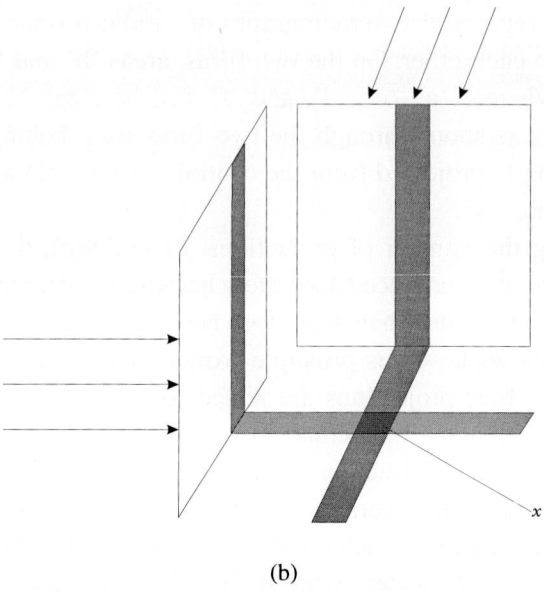

(b)

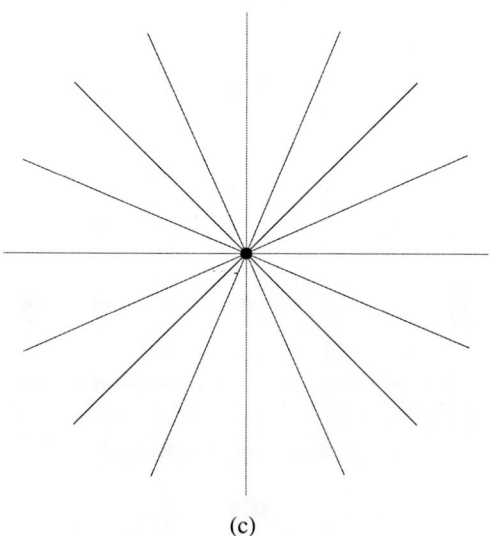

(c)

Fig. 2.9 (*Continued*)

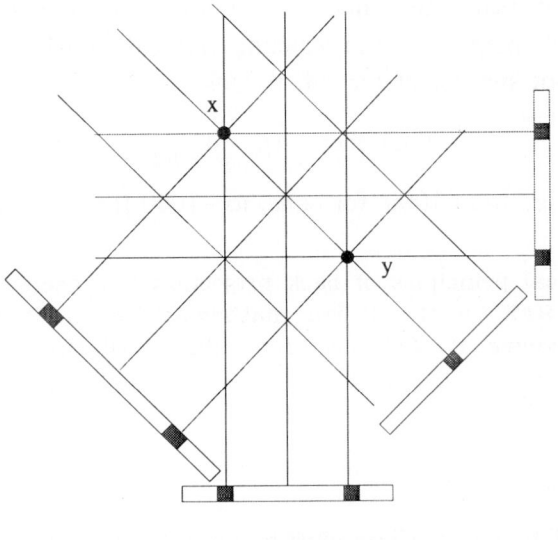

(a)

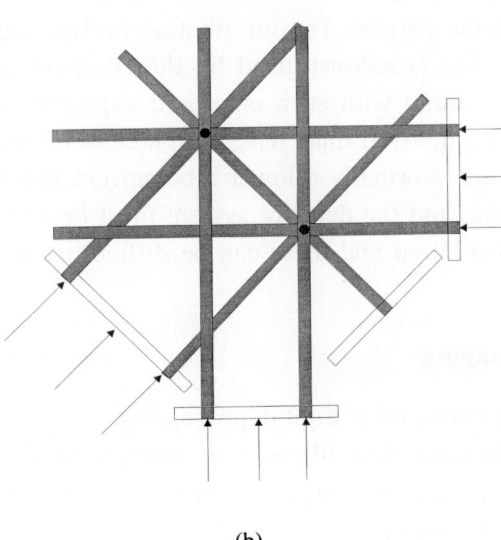

(b)

Fig. 2.10

m_{tissue} are not necessarily the same as each other, CT numbers vary slightly over the range of energies used in scanning, (typically 90-140 kV), but these variations are not significant for soft tissues.

$$CT \text{ number} = (\mu_t - \mu_w)\mu_w^{-1}$$

CT numbers range from 1000, for bone, to −1000 for air, with 0 for water.

The mathematical principles of back projection had been formulated as long ago as 1917 by the Polish mathematician, J. Radon, working in Vienna. However, it was not practicable until the advent of fast computers.

Spiral Scanning

With normal CT scanning, slices must be produced sequentially. The couch is moved between each tube rotation. For spiral scanning, the couch moves steadily and continuously as the tube and detectors make a number of revolutions round the patient. The information is thus acquired in a series of slanting slices, but is reconstructed by the computer as vertical slices. These are easier to align with each other and exposure time is lower.

The X-ray tube employed must have a high capacity because of the long continuous exposure. Normally a lower tube current still has to be used to prevent overheating, and the detector system must be very sensitive. There is some loss of resolution and there can be difficulties with large patients.

Digital X-ray Imaging

Computed tomography uses digital processing as do other some other investigative techniques like ultrasound, nuclear medicine imaging and magnetic resonance imaging. There are advantages for X-ray imaging too of digital image reconstruction.

1. Image processing can improve contrast and other aspects of image quality to produce superior pictures.

2. Radiographs can more easily be compared with pictures from other imaging modalities.
3. Images can be distributed electronically within hospitals, and remote accessing and archiving are possible.
4. Highly qualified personnel can access images from remote locations at a central base.
5. Space may be saved because it is not necessary to keep bulky hard copies.

The Production of Digital X-ray Images

As has been demonstrated, analogue X-rays are produced using a film cassette containing screens. These absorb X-rays and give off light to provide the image on the film.

There are three different approaches to the production of digital images.

1. The signal from a video camera which is optically coupled to an image intensifier can be digitised to produce an instant read out.
2. This method also uses a phosphor screen, but of a material which contains traps for electrons excited by incident X-rays. By illuminating the phosphor with red laser light, the latent image formed by trapped electrons is extracted in the form of a blue light image. This does not produce instant images because the cassette has to be scanned by the laser and read out by a photomultiplier to digitise it.
3. This is based on the principle, originally discovered in 1902 by Fürstenan in Germany, that the element selenium changes its electrical resistance when irradiated with X-rays. X-ray photons are converted directly into charge carriers using a sheet of amorphous selenium photoconductor (a-Se). This is also the principle used in xerography. After exposure, the image appears as a charge distribution on the a-Se surface. This can be read out electrostatically and then digitised.

The first two methods have the disadvantage of rather poor resolution because light has to be collected by a thick enough layer to stop the energetic X-rays.

The drawback, at present, of the third method is that the equipment is bulky and it cannot be housed in most existing X-ray rooms. It employs a very large rotating drum whose surface must be kept exactly 0.1 mm from the electrometers which measure the charge.

In newer, purpose built, departments of radiology there may be fewer problems and researchers are working on new detector systems using the same amorphous selenium but flat panel detectors. (3)

Radiographic Subtraction

This technique involves the superposition of two radiographs of the same region which contain different information. One is subtracted from the other so that only information which is different between the two pictures is shown. Digital radiography makes this process easy. The purpose is to be able to eliminate anatomical details and show only contrast medium in structures such as blood vessels. It is possible to see more information with the administration of less contrast medium.

Images can be spoiled by patient movement due to breathing, swallowing or vascular pulsation, as well as involuntary moves. With digital techniques motion artefacts can often be eliminated by realigning the images pixel by pixel.

Energy subtraction can produce similar subtraction information. Two exposures, one with high kV and the second with a much lower kV can be made rapidly before any patient movement is made. The two images will have very different bone to soft tissue contrast and can be subtracted from one another in different weightings to show either bone or soft tissue alone.

Bone Mineral Density Measurement

Dual energy X-ray absorptiometry (DEXA) is an established method of measuring bone mineral density. This is an important indicator of the presence, or likelihood, of osteoporosis.

A specially designed X-ray unit is required, consisting of a highly stable X-ray tube producing a narrow fan, or pencil, beam of X-rays which can be scanned across the patient. Usually it is the vertebrae that are examined. After passing through the patient the transmitted radiation is collected by a detector.

The energy spectrum of the X-rays has two distinct peaks; a high energy part produced from high voltage applied to the tube, and a lower energy region achieved by passing the beam through a k-edge filter. Measurement of the total radiation transmitted for each of the two spectral regions allows the mineral composition to be calculated, (5). Studies must be repeated on an annual basis for vulnerable patients, but the patient dose is low because of the restricted beam size.

Osteoporosis is a growing problem because of an increasing population of older people. It leads to reduced mobility, deformity, and the development of fractures. Early diagnosis helps to establish treatment which can slow the progress of bone demineralisation.

References

1. Freund L., *Elements of General Radio-Therapy for Practitioners*, Vienna, 1903.
2. White D. R., Peaple L. H. J. and Crosby T. J., *Radiat. Res.* **84**, 239–252, 1980.
3. Rowlands J. and Kasap S., *Physics Today* **50,** 24–30, 1997.
4. Pusey W. A. and Caldwell E. W., *The Röntgen Rays in Therapeutics and Diagnosis*, W. B. Saunders, 1903.
5. Fogelman I. and Blake G. M., *Current Research in Osteoporosis and Bone Mineral Measurement III*, London, British Institute of Radiology, 1994.

Chapter 3

TREATMENT WITH
IONISING RADIATION

Throughout history, members of the medical professions have tried to use all the available technologies for the treatment of disease. Before the discovery of X-rays and radioactivity, heat and light had both been applied enthusiastically to treat various conditions. A physician named Muravka in Russia had even tried the application of radiation from glow worms in the early nineteenth century.

In 1893, Nikola Tesla, a Jugoslavian physicist who had emigrated to America, produced rapidly alternating electric currents which gave brilliant lighting effects when passed through partially evacuated glass tubes. Tesla currents were found to produce powerful physiological changes in small experimental animals, and were subsequently applied to human patients. An 1893 French opinion was that Tesla currents *increased oxygenation of the body, promoting the elimination of uric and phosphoric acid.* Other beneficial effects were also proposed. In 1899, newspapers published sensational reports that high frequency currents had succeeded in curing pulmonary tuberculosis.

Figure 3.1 shows a large solenoid called a *Cage of Autoconduction*, seven feet long and three or four feet in diameter with twenty or so turns of stout copper wire. This was sold to medical people and it was claimed to be successful in the treatment of various conditions from nutritional deficiencies to rheumatism and insomnia.

The therapeutic use of electrical and magnetic fields, and of ultra-violet light, visible light and heat, were classed as 'Radiotherapy', and this is why they are mentioned here, although today, radiotherapy means treatment with ionising radiations.

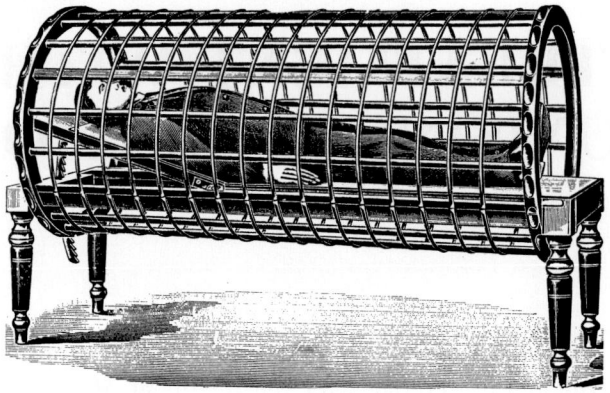

Fig. 3.1. Horizontal cage. (*Sanitas Electric Co.*, London.)

It was into this environment of empirical and unexplained remedies that X-rays and radioactive sources were first introduced shortly after their discovery. The first published textbook of the twentieth century in this field, entitled *Elements of General Radiotherapy* by Dr Leopold Freund in Vienna in 1903, included all the foregoing forms of radiation. Non-ionising radiations will be covered in the next chapter.

The Development of Radiotherapy

The first attempt to use X-rays for treatment, made only six months after their discovery, was by Dr Leopold Freud in Vienna. He had heard that a man working with X-rays had contracted a dermatitis accompanied by hair loss. He then read a report of similar hair loss in a young man in Berlin whose head had been irradiated. He had, at that time, a young female patient with a large hairy naevus on her neck and back which he treated as follows:

'A Ruhmkorff's coil by Keiser & Schmidt of Berlin, having a 25 cm spark length, was worked from accumulators, and vacuum tubes which had been proved to be rich in X-ray powers were used. The tubes were capable of giving a Röntgen photograph of a man's hand at a distance of 15 cm with one minute's exposure, and the fluorescent screen was used from time to time to prove their proper working. These tubes possessed platinum anti-cathodes and aluminium electrodes. ... The child was made to sit with her back bared, and the tube was placed at a distance of 10 cm, so that the zone richest in X-rays coincided with the nape of the neck. In this way a large part of the naevus from the scalp down to about the middle of the dorsal spine was irradiated. At first a thick leaden mask with an aperture corresponding to the area of the naevus was used; this, however, was soon dispensed with, the result of the experiment being apparently so doubtful. The exposure was two hours daily.

The author soon convinced himself that no perceptible heat was evolved from the tube. The child, though a sensitive one, endured the sittings extremely well and remained well and lively throughout. For the first ten days no change whatever occurred; not a single hair fell. On the eleventh day the mother removed several bundles of loose hairs from the inter-scapular region, and the author did likewise. With gentle pulling, bundles of 5–10 hairs came out each time in one's fingers, with absolutely no sensation of pain on the part of the child. ... Eight days after the commencement of hair shedding, a dermatitis developed from two small excoriations which had already been noted on the nape. ... The dermatitis disappeared after a few days treatment with ichthyol ointment, and with it went the few remaining hairs at the nape.' (1).

Unfortunately, the patient subsequently developed a very severe skin reaction and her general health began to deteriorate. She was in considerable pain and lost her appetite. The irradiated area developed ulceration which did not begin to heal for several months. When seen four years later, there was an unpleasant and unsightly scar which itself gave way as a result of slight trauma. Some attempts were made to treat this surgically.

Not surprisingly, this patient had been over-dosed, partly because the experimenters did not know that the effect of treatment would not be immediate.

In August 1896, Despeignes in Lyons, France, reported that 'a case of gastric carcinoma appeared to have been greatly benefited by the transmission of X-rays through the seat of disease.' (2).

Freund and Schiff followed the first experiments in Vienna by treating acne, lupus erythematosis, and other skin diseases. Later they treated cancer of the pharynx, in 1898, and were successful in relieving the pain.

Fig. 3.2

In England, Johnson and Merrill reported favourable results in skin cancer following X-ray treatment in 1900, and, at about the same time, Clark managed to shrink cancerous ulceration in breast cancer (3).

By 1903, a radiotherapy department had been established at the London Hospital, Whitechapel, (Fig. 3.2), and a great variety of different forms of cancer had received treatment. In these early days radiotherapy was prescribed as often for non-malignant conditions as for cancers. In addition to the treatments mentioned above, ringworm, tuberculous and osteomyelitic joints, ankylosing spondilitis and thyrotoxicosis responded well, but later it was found that there was a substantial risk of the development of cancer in such patients. Smaller radiation doses were still being given to stimulate the pituitary gland or bring about an artificial menopause into the 1960s. After that, radiotherapy for most non-malignant conditions was discontinued.

At first, treatment doses were usually specified in terms of length of time of exposure, and it became necessary, for consistency and to try to establish curative doses but eliminate overdosage, to devise means of measuring the actual quantities of radiation being delivered.

The Action of Radiation on Tissue

Irradiation causes ionisation within cells which can lead to cell death because of damage to the DNA molecules. The interaction with cell processes is extremely complex, and the reader is referred to various works on radiation biology for fuller explanation, (4), but the effectiveness of radiotherapy treatment depends upon an understanding of radiation damage.

The average energy expenditure per ion pair formed in air by X-rays is approximately 34 eV. About half of this is due to ionisation and half to excitation in air atoms. The time scale of the physical interaction resulting in an ionisation event is about 10^{-13} s and, in human tissue, this initial event is followed by a chemical phase and then a biological one.

During the chemical phase of about 10^{-9} s, short lived free radicals are produced which interact with each other and with the surrounding tissue. The effect is enhanced by the presence of oxygen. The following biological phase, lasting some 10^{-6} s, allows the radicals and ions formed to interact with constituent parts of cells causing some biological changes in the structure or function of the cell. Gross damage occurs during cell mitosis and this eventually kills the cell even though it may be capable of a few further divisions.

Ionisation occurs in all living tissue, both normal and abnormal, and the success of treatment depends upon recovery of normal tissues and destruction of cancer cells. There will always be a greater number of cancer cells undergoing mitotic division than normal cells undergoing division, and therefore a greater probability of killing cancer cells. Radiotherapy must be delivered with great geometrical accuracy and dose fractionation must be correct.

The radiotherapy physicist has three main functions:

1. To ensure that the doses delivered are accurate.
2. To decide on the best combination of radiation fields which will give the highest dose to the tumour volume and the lowest possible to surrounding normal tissue.
3. To advise on appropriate shielding around radiotherapy installations, and make any necessary radiation protection measurements from time to time.

Radiation Measurement

The first attempt by a physicist to measure quantity of ionising radiation was made in 1897 by Dorn in Würzburg using an air thermometer. Rutherford and McClung in England in 1901, used a bolometer to try to determine the total energy emitted by both a röntgen tube and a radium source.

Following the realisation of the greatly delayed reaction in patients who were irradiated, it became more important to find some reliable means of dose estimation.

A scheme proposed first by Holzknecht in Germany in 1902 was based on the principle that certain chemicals changed colour when irradiated. Holzknecht's device, which he called a chromoradiometer, was a series of capsules containing a solution whose composition he did not divulge. It was not satisfactory because of variable results.

In 1904, Sabouraud in Paris working with Noiré produced a colour radiometer containing barium platinocyanide. This was green when unexposed but turned to orange when a fairly high dose of radiation had been reached. It was said that, if the detector was positioned half way between the X-ray tube and the patient's skin, the orange colour appeared when a dose sufficient to cause the hair to fall out — (the epilating dose) — had been reached.

Pastille dosemeters based on colour changes were then developed. A colour chart was issued with the dosemeters showing a graduated scale running from colourless through orange to deep brown with increasing dose.

Efforts were also made to use fluorescence as a measure of dose, while Stern in New York tried photographic blackening. This led to the development in Vienna in 1904, of the Kienböck Quantimeter which was a series of photographic paper strips coated with silver bromide. A grey-scale of degrees of blackening corresponding to an arbitrary scale was provided for comparison.

Another German, Fürstenan, discovered in 1902 that the electrical resistance of metallic selenium changed when it was irradiated with light or with X-rays, and an intensimeter based on this principle was introduced in 1915. This was the first instrument to measure units which were called 'I' units by the inventor. This was also the first practical instrument which could measure the output of an X-ray tube while it was in operation.

The selenium strip was enclosed in a light-tight packet which could be placed on the skin of a patient receiving therapy. Electrical leads were connected to both ends of the metal strip and enclosed in a long cable and this acted as one arm of a Wheatstone Bridge. A box containing the rest of the circuitry and a galvanometer was placed some distance away. The decrease in resistance of the selenium when irradiated unbalanced the bridge, and the degree of imbalance, which was proportional to the radiation intensity, could be read from the galvanometer.

The introduction of the Coolidge tube supplied by a closed core transformer led to the development of X-ray equipment with a steady output making it possible to control voltage, current and exposure time. By about 1916, epilation and erythema doses were being specified by voltage, current, target to skin distance, filter, and exposure time.

Ionisation

Röntgen had discovered ionisation a few weeks after discovering X-rays, when he noted a change in electrical conductivity when matter was irradiated.

Electroscopes and electrometers were used in the 1890s by Wilson and others in England, and by Geitel in Germany. In 1905 an article by M. Franklin appeared in the New York Medical Journal describing the disadvantages of both photographic and pastille methods and describing the use of the electroscope. (5)

An electroscope consists of a metallic box containing an aluminium fibre mounted on a metal post, and an electrostatic charger. The fibre discharges when irradiated and the degree of discharge is proportional to the radiation dose.

A new unit of exposure was proposed in 1908 in France by M. P. Villard to be: 'The quantity of X-rays which liberates one electrostatic unit of charge per cubic centimetre of air at normal temperature and pressure.' This was called the 'e' unit.

In 1930, the International unit of exposure, the Röntgen, was defined as "the quantity of X-radiation which produces in 1 cubic centimetre of atmospheric air at 0°C and 760 mm mercury pressure, 1 esu of charge."

Marie Curie realised that the radiations discovered by Becquerel could also be measured by their ionising effect. She built the Curie Electrometer which was equipped with an ionisation chamber, and she used it to demonstrate radiation intensity emitted from uranium and thorium.

Curie wanted to establish a unit of activity for radium and the other radioactive elements which she and others were discovering. By now occupying the chair of physics in Paris which her husband had held until his death in 1908, she was a prime mover of the Radiology Congress held in Brussels in 1910 which defined standards for radium and its dosimetry. The Congress defined the "curie" as the unit of radioactivity as the 'quantity of radon which is in radioactive equilibrium with 1 gram of radium'.

This was only useful for radium, where the number of nuclear transformations per second is 3.7×10^{10}. The definition was later extended to include all radioactive isotopes as 'a unit of activity which gives 3.7×10^{10} disintegrations per second.'

In 1923, the Röntgen Society and the Physical Society jointly set up the British X-ray Units Committee under the chairmanship of Sir William Bragg. While attempting to introduce uniformity into radiation measurements of exposure, activity and absorbed dose, this committee was asked by the First International Congress of Radiology held in 1925, to set up an International X-ray Units Committee. From these beginnings there emerged the British Committee for Radiological Units (BCRU) and the International Committee for Radiological Units (ICRU), both of which still exist.

Dose Measurement by Ionisation

A useful measuring system must be reproducible, sensitive and linear. It must be independent of the intensity of the radiation, but capable of measuring both large and small doses. It should also operate over a wide energy range.

A gas filled ionisation chamber can easily be made to fulfil these requirements. The 'thimble' or 'cavity' ionisation chamber, as shown in Fig. 3.3, consists of a plastic thimble filled with air and with a thin metallic central electrode. A polarising voltage is applied between the outer thimble wall and the electrode.

An X-ray photon absorbed in the wall liberates a secondary electron which produces ion pairs along its track. Because of the polarising voltage, these ions are separated before they have a chance to recombine. The positive ions are attracted to the chamber wall (negative electrode), and the negative ions to the central electrode. An ionisation current flows, proportional to the dose rate of the radiation and the mass of air in the chamber.

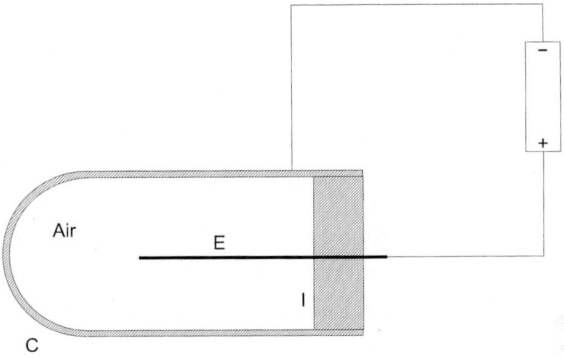

Fig. 3.3. Thimble Ionisation chamber: C = Cap, E = Central electrode, and I = Insulator.

A polarising voltage of about 100 V is normally sufficient so that all ions are collected. A voltage which is too low will allow some of the ion pairs to recombine, but, above a certain saturation voltage, the ionisation current will be constant regardless of the polarising voltage.

Dose may also be measured electronically by means of the photo-conductivity of germanium or silicon. This dosemeter consists of a small semiconductor crystal in series with a battery and a digital measuring device. Radiation absorbed by the crystal makes it conduct a current which is proportional to dose rate. There are several types of semiconductor detector, which are selected to fit best the radiation under investigation.

Other methods of dose measurement are by thermoluminescent dosimetry and by scintillation counters (which will be described in the chapter on nuclear medicine).

Thermoluminescent Dosimetry (TLD)

Radiophotoluminescence is the emission of electromagnetic radiation from a substance due to causes other than heating it to incandescence. As discussed in Chap. 2, fluorescent and phosphorescent materials respond to ionising radiation by the raising of valence electrons to higher energy levels. The electrons drop back spontaneously into their valence shells with the emission of light. Instantaneous reversion to valence shells is called fluorescence, but where there is a time interval, the phenomenon is called phosphorescence.

Thermoluminescent materials are affected in the same way, in that electrons are excited to higher levels by ionising radiation, but they do not drop back into valence shells, neither is light emitted, unless and until the material is heated to about 300°C. Carefully controlled heating causes the trapped electrons to return to their ground state with the emission of light, and the total amount of light is proportional to the radiation received. Calibration is easy by exposing TLD chips to known amounts of radiation and processing them with the experimental chips.

TLD is extremely effective and useful. The materials involved are stable and the process of read-out can be delayed for a long time. The action is largely independent of radiation energy and very low doses can be measured. This makes TLD the method of choice for personnel monitoring now that new radiation protection regulations require the recording of doses which are too low for the older film badge dosimetry. It is also important for the measurement of patient doses in radiotherapy, being quicker, easier and cheaper than other methods, and TLD dosemeters are small and tissue equivalent so that patient's treatment doses are unaffected by their use. Their small size makes it possible to mount TLD chips in small sachets that can be fixed on any part of the body to measure personnel extremity doses.

When TLD dosemeters have been processed, they can be heated further to remove all the deeply trapped electrons from higher shells and, having returned to their pre-irradiated state, they can be reused. Automatic TLD readers are available which can read a bar code on a well designed card containing TLD chips, carry out the annealing cycle, calculate the total light emission and record the dose results.

There are several common materials which have thermoluminescent properties. Those most commonly used in medicine are of lithium fluoride because of its low energy dependence and good approximation to human tissue. Calcium sulphate is used when a particularly low doses are to be measured.

The reader is referred to Refs. (6) and (7) for fuller information on TLD.

S.I. Units of Radiation Measurement

The units in use for radiation measurement now are as follows:

Absorbed Dose

Since exposure does not describe the energy imparted to a material which is irradiated, it cannot be used to specify radiation energy absorbed. The unit of absorbed dose before the introduction of S.I. Units was the rad where 1 rad was 100 ergs per gram. The S.I. Unit is the gray (after the British physicist Louis Harold Gray, 1905–65). 1 gray = 1 joule per kilogram.

Because of the large size of this unit, it is often useful to use the prefixes

$$m = micro = 10^{-6}$$
$$m = milli = 10^{-3}$$
$$c = centi = 10^{-2}$$

Activity

The S.I. unit of activity is the becquerel (Bq). 1 becquerel = 1 disintegration per second. Thus 1 curie = 37×10^9 Bq.

Because of the very small size of this unit the following prefixes are common:

$$k = kilo = 10^3$$
$$M = mega = 10^6$$
$$G = giga = 10^9$$
$$T = tera = 10^{12}$$

Dose Equivalent

Dose equivalent takes account of the different damaging power of different radiations. This concept was formerly covered by RBE or relative biological effectiveness. The unit of dose equivalent is the sievert (after the Swedish physicist Rolf Sievert, 1896–1966). Its numerical value is the same as absorbed dose for X- and gamma rays, but a quality factor is required for other types of radiation.

Radiotherapy Treatment Machines

As has been described in Chap. 1, X-rays below about 0.1 MeV are useful for diagnosis, but, because of the high degree of absorption in body tissues, they are not sufficiently penetrating to be useful in radiotherapy.

Following the end of hostilities in World War 1, news emerged that the Germans had developed X-ray tubes capable of functioning at over 200 keV, and this led to the mass production of such tubes for what was called *deep X-ray therapy*. The classifications now used for external beam therapy, determined by the generating voltage, are:

Superficial therapy = < 200 keV
Orthovoltage therapy = 200–400 keV
Megavoltage therapy = > 1 MeV
(X-rays above Co-60 quality are often referred to as supervoltage)

Additionally, electrons are used for external beam treatment.

Orthovoltage Therapy

There are differences between the X-ray tubes used for diagnosis and for therapy. Electrons from a heated filament are accelerated across an evacuated tube in both cases, but, for orthovoltage therapy a high voltage generator is required to supply the appropriate voltage and current, and the tube must be designed to withstand that voltage and to cope with the heat generated by the production of higher energy X-rays.

Cooling is usually achieved by a flow of oil over the back of the anode, and the hot oil is passed into a cooling radiator where it gives up its heat either to surrounding air or to water. In diagnostic radiology the size of the focal spot must be very small to provide a sharp image. In radiotherapy this is not so important and focal spots of about 1 cm diameter

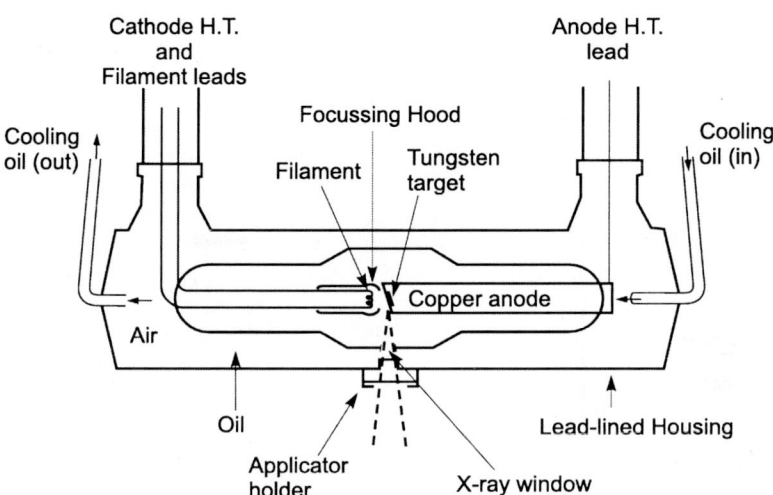

Fig. 3.4

are common.   The target angle in a diagnostic tube is usually 17° but for radiotherapy it is 30°. This gives the most uniform beam at 50 cm focus-to-skin distance.

Figure 3.4 is a diagrammatic cross section through a typical orthovoltage X-ray tube.

For superficial and orthovoltage therapy, the radiation beam is collimated by a metal cone or applicator. Applicators of various common sizes are supplied with the treatment machine and it is possible to cope with odd shaped fields by the employment of 'cut-out' shapes in lead. The more energetic the radiation, the thicker the lead.

Figure 3.5 shows how radiation intensity falls as the beam passes through a patient. At orthovoltage quality the beam is down to 50% of the value at the skin surface at a depth of 6 cm, and although it is possible to combine beams of radiation at various angles to increase tumour dose, it is still only possible effectively to treat fairly superficial cancers with this energy of radiation. Figure 3.6. shows isodose curves for a 5 cm × 5 cm field at 250 kV and for a cobalt-60 gamma ray beam.

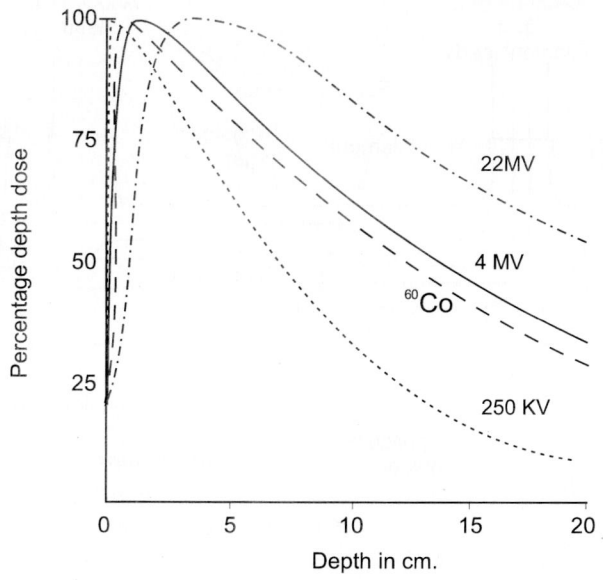

Fig. 3.5

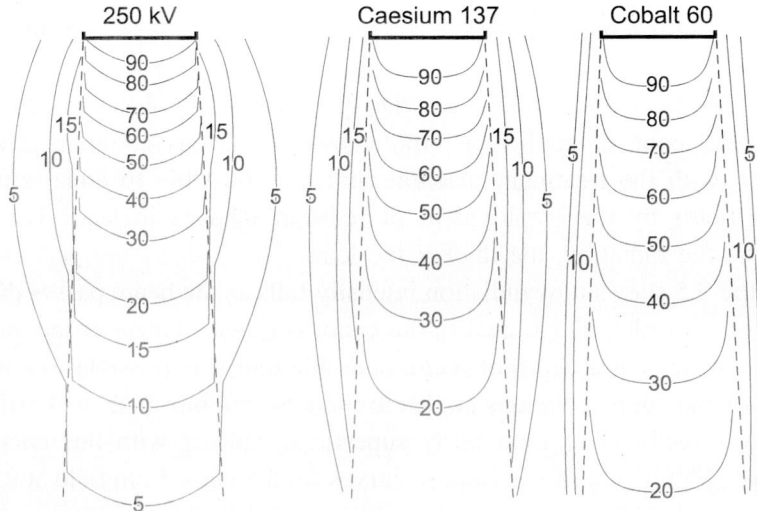

Fig. 3.6

Cobalt-60 Radiotherapy

Cobalt-60 is an artificially produced isotope which became available commercially when nuclear reactors became common. Nuclear reactors were established all over the world for the production of weapons following the development of the first atomic bombs dropped on Hiroshima and Nagasaki in the second world war. The fission process in reactors produces large quantities of different radioactive materials with a great range of different energies and radioactive lifetimes.

The first use of a radioactive substance to generate a beam of gamma rays for treatment (called teletherapy) involved radium. Four radium 'bombs' were built in the 1930s by physicist Henry Flint for four London hospitals; the Cancer Hospital (now called the Royal Marsden Hospital), the Middlesex Hospital, University College Hospital and the Westminster Hospital. Each was a simple construction of a sphere of lead, hollowed out to contain a pack of radium. A lead stopper made the machine reasonably safe when not in use, and the stopper was remotely operated by a long cable following positioning against the lesion on the patient.

Radium, with its gamma energy of 2.4 MeV, presented radiation protection problems even before the introduction of radiation dose limits for staff. Additionally radium has a troublesome gaseous radioactive daughter product, radon, which builds up inside sealed source containers. Cobalt-60 has a half life of 5.2 years and emits gamma rays of 1.17 and 1.33 MeV, thus its penetrating power is superior to orthovoltage X-rays although less powerful than radium.

Megavoltage therapy has other advantages. Above 1.0 MeV there is an equivalent absorption of energy per unit mass in most tissue materials. Bone, fat and muscle all absorb approximately equal amounts because the main absorption process is due to the Compton effect which is independent of atomic number. The second advantage is a surface effect where the radiation beam enters the body. Figure 3.5 shows that, at 250 keV, the maximum radiation dose occurs in the surface of the skin. This causes severe skin reactions when radiotherapy is given. At cobalt quality and above, the maximum is below the surface due to the gradual build up of secondary electrons to an equilibrium situation. These are mostly Compton recoil electrons which are liberated by the interaction of the radiation with the superficial tissues. Radiotherapy machines employing cobalt were much in demand from the early 1960s.

At about the sause time, another fission product, caesium-137 was thought suitable for external beam therapy. The gamma ray energy of caesium-137 is 660 MeV so the protection requirements were less complicated than for cobalt-60, and the half life of 30 years was more convenient for dosimetry. Caesium-137 units became popular in smaller radiotherapy centres before the greater penetrative power of cobalt-60 assumed more importance.

A typical cobalt-60 treatment unit is illustrated diagrammatically in Fig. 3.7. The source is a sealed cylinder measuring about 2 cm long and 1.5 cm diameter containing a number of thin discs of cobalt-60 totalling about 200 terabecquerels. This is mounted near to the edge of a circular drum

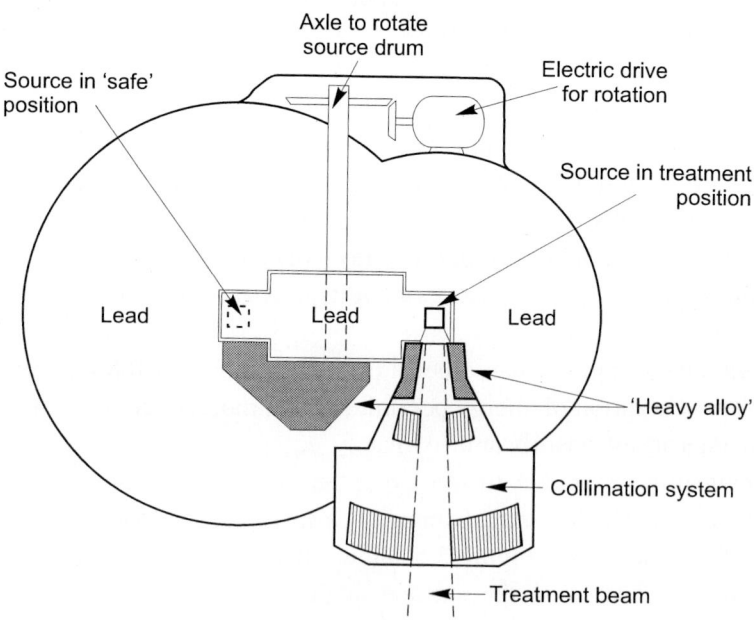

Fig. 3.7

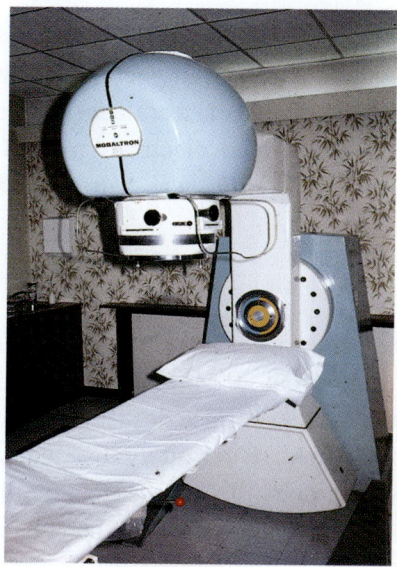

Fig. 3.8. Telecobalt machine (Mobaltron). Courtesy of TEM.

pivoted on the common axis of the two spheres which make up the head of the treatment machine. The machine is switched on by causing the drum to rotate through 180° so that the source is above the beam port through which the treatment beam emerges via collimators. In all other directions the gamma rays are cut off by the surrounding lead.

The treatment head is mounted so that it can rotate around the treatment couch as shown in Fig. 3.8. This allows isocentric treatment to be administered with all radiation fields centred on the centre of the tumour volume.

Supervoltage Therapy

An early attempt to produce X-rays of over 1 MeV was the Van de Graaff electrostatic generator in 1931. This was developed by Robert Van de Graaff at Massachusetts Institute of Technology (8), (9). 1 MeV X-ray therapy was used at a number of radiotherapy centres from the mid-1930s, but the equipment was cumbersome and expensive.

In 1940, Donald W. Kerst at the University of Illinois designed the betatron. This consisted of a reasonably compact annular ring in which electrons, guided by a magnetic field, were accelerated in circular orbits. A medical version, which accelerated electrons to 2 MeV, was designed in 1943, but progress of this technology was halted by the second world war.

Linear Accelerators

Linear accelerators are used to generate treatment beams of X-rays above cobalt-60 quality and up to about 20 MeV. Above that energy, although there is less scattered radiation, smaller surface dose and ever increasing penetration, there are various factors which are definitely disadvantageous. The shielding effect of bone increases as does the excess dose within the living parts of bone. There are also high transmitted doses increasing the chance of skin reaction on the exit surface of the patient.

A linear accelerator is illustrated diagrammatically in Fig. 3.9. Electrons, travelling at about half the speed of light, are injected into an evacuated tube where microwaves are used to accelerate them to about 0.95 times the speed of light. These high energy electrons bombard a transmission target of tungsten or tungsten-copper alloy, to produce X-rays in the range 4–20 MeV. A metallic beam flattening filter is required to provide a suitable beam for treatment purposes because the X-ray beam is concentrated towards the centre line. The higher the energy, the thicker is the centre of the flattening filter. Thus, another disadvantage of very high energies is output reduction due to the size of this filter.

Linacs within the range 4–20 MeV are capable of delivering much higher dose rates than cobalt-60 machines, making it possible to treat many more patients per day and increasing cost effectiveness.

Microwave Sources

The development of the linear accelerator was made possible by wartime work on radar in the 1940s which led to the production of high power, high frequency microwave sources. These pulsed diode sources, the magnetron

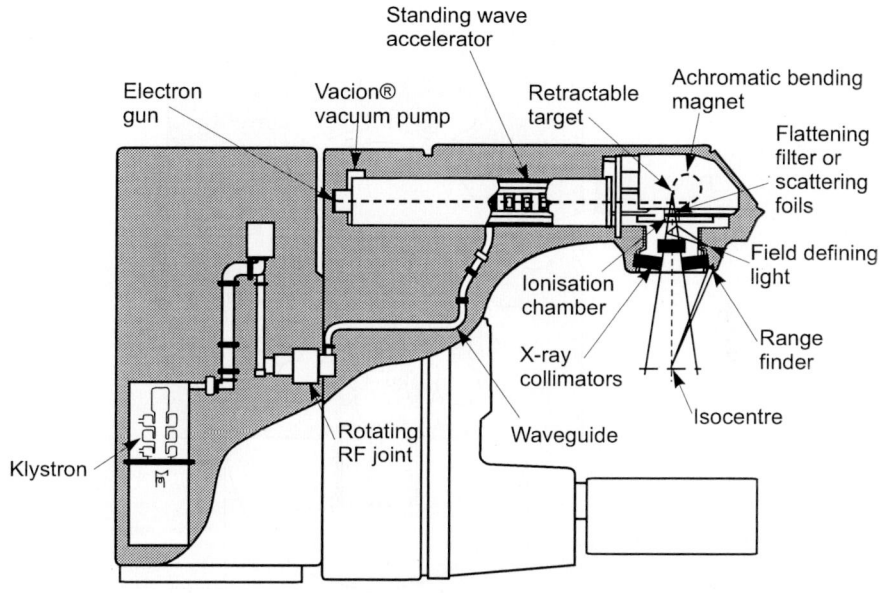

Fig. 3.9. (Diagram courtesy of Varian Associates USA).

and the klystron, can establish intense electromagnetic fields in suitably designed cavities. A magnetron is smaller and cheaper than a klystron and can be mounted on the linac's gantry drum, but it does have a considerably shorter useful lifetime than a klystron.

A magnetron, operating at a peak power of 2.5 MW, is useful for electron energies up to about 10 MeV. For higher energies, magnetrons or klystrons operating at a peak power of 5 MW are required.

A magnetron consists of a cylindrical heated cathode enclosed in a cylindrical anode containing resonant cavities, as shown in Fig. 3.10. This is placed in a uniform magnetic field whose lines of force are normal to the plane of the figure. The magnetic field strength and anode potential are arranged so that electrons ejected from the cathode travel in curved paths, just failing to strike the anode and returning towards the cathode. Oscillations are set up in the anode cavities and the resulting microwaves or R-F are extracted by means of a coupling loop.

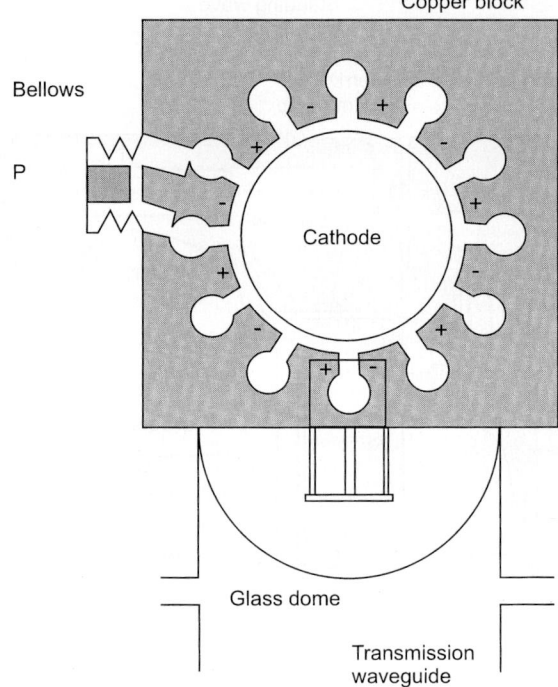

Fig. 3.10

In the klystron, (Fig. 3.11), a low powered oscillator excites the primary resonant cavity. Electrons projected from the cathode into a small gap between the cavities are either accelerated or decelerated depending on the phase of the electric field. Electrons in the early phase travel more slowly than later ones which overtake them, so that the electrons arrive downstream in bunches at a frequency determined by the resonant frequency of the first cavity.

The Waveguide

The electron gun is pulsed so that electrons are injected in bunches into the waveguide, in synchrony with the R-F pulses from the magnetron or klystron. The electrons are focused electrostatically onto the central axis of the

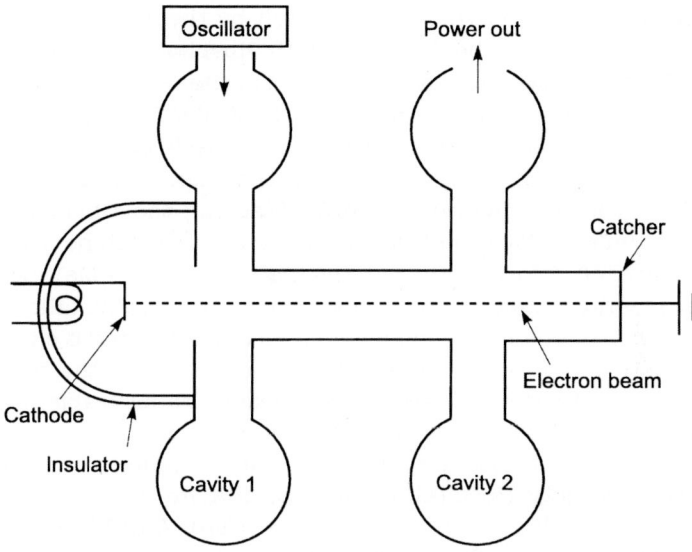

Fig. 3.11

waveguide by steering coils aligned along the outside of the waveguide. In addition to preventing divergence, these reduce the generation of unwanted bremsstrahlung. Being negatively charged, the electrons are accelerated during half of the R-F cycle.

There are two kinds of waveguide. A travelling waveguide is a series of cylindrical, evacuated microwave cavities. The spacing between cavities increases along the guide. A progressive R-F wave with controlled phase velocity moves along the waveguide providing acceleration to the proportion of electrons 'captured' by the correct half cycle. In a standing waveguide the R-F wave is reflected back down the guide after passing through a phase adjuster. The result is a spatially stationary wave that is oscillating in time.

Standing waveguides are able to produce higher accelerating gradients, leading to more compact treatment machines, but at the expense of an increased energy spread of the accelerated electrons, and this can cause problems at the bending magnet, or can lead to an inhomogeneous radiation beam.

The first medical linear accelerator was built by the engineering company Metropolitan-Vickers for the Hammersmith Hospital, London in 1953. It generated 8 MeV X-rays and had a fixed horizontal structure with a three metre long waveguide. Shortly afterwards, Mullard provided a 15 MeV linear accelerator with a 6 metre waveguide capable of delivering a beam of either electrons or X-rays at St Bartholomew's Hospital, London. Metropolitan-Vickers provided a 4 MeV linac for the Christie Hospital, Manchester, while Mullard built the first isocentric machine, capable of a reasonable arc of rotation around the patient, with 4 MeV X-rays for Newcastle General Hospital. These three machines were installed in 1954. The Newcastle machine had a one metre waveguide making it cumbersome, and in order to allow 105° rotation from the vertical of the head, it was necessary for the floor of the treatment room to be moveable downwards to allow treatment from below the horizontal.

The first machine to use a klystron as the R-F power source was built at Stanford University Hospital, California, USA in 1955.

By the mid 1990s there were more than 4000 medical linear accelerators in use throughout the world.

The technique employed to provide isocentric linear accelerators capable of full rotation round the patient with accessible isocentre height, was to mount the wave guide horizontally and use magnetic fields to bend the electron stream at the end of the accelerating waveguide before it hits the target. The first such machines used a 90° bending magnet system, but a 270° bending magnet was subsequently found to have advantages in providing improved stability and beam symmetry.

Electron Radiotherapy

The first machines to produce electrons for treatment purposes were betatrons which were large and expensive. Few betatrons remain in use in Radiotherapy Departments. Electrons which have been accelerated along the waveguide of a linear accelerator are now normally used as treatment beams with the X-ray target removed.

The stream of electrons from the waveguide contains a range of energies and this range can be altered by adjustment of the microwave power from the magnetron or klystron. Electron beam energy is approximately proportional

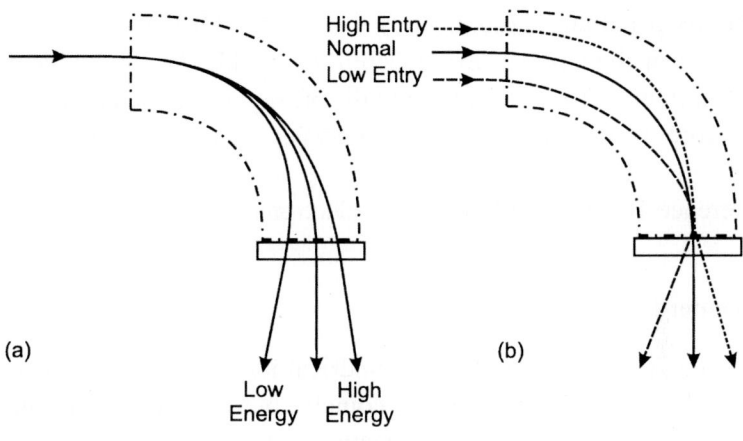

High Entry
Normal
Low Entry

Low   High
Energy   Energy

(a)   (b)

Fig. 3.12

to the square of the delivered microwave power pulse. Electrons of the required energy are selected by the bending magnet. Figure 3.12 shows how this is done with both 90° and 270° magnets.

The electron beam emerging from the bending system has only a small cross sectional area of diameter a few millimetres. It can be converted into a widely spread beam covering a large surface area evenly by one of three methods:

1. The provision of a thin foil of high atomic number material.
2. The provision of a pair of scattering foils.
3. Magnetic scanning of the beam.

A single foil gives minimum energy loss during transmission and low radiative loss. This avoids contamination of the beam by bremsstrahlung. A double foil gives a much flatter electron beam but with decreased intensity. Scanning magnets, operating in a similar fashion to those used in the production of a television picture, overcome the problems of both bremsstrahlung and energy loss, but at the expense of producing instantaneous high local point intensity, making beam measurement extremely difficult. All three methods have been used on electron linacs, but it is now recognised that dual scattering foils produce the best compromise solution.

Electrons penetrate approximately 1 cm per 2 MeV so useful treatment depths between about 5–10 cm are therefore possible. In spite of the presence of bremsstrahlung, the dose in tissue falls off rapidly at the end of the range, and therefore the dose to normal tissue deeper than the treatment volume is minimal.

Reference 10 gives full details of electrons for treatment purposes.

Measurement

Electrons can be measured with a cylindrical ion chamber, but the effective measuring point does not necessarily coincide with the position of the central electrode of such a chamber. Exact position of the measuring point is usually required in radiotherapy dosimetry, so it is more usual to use a flat, parallel plate chamber except at higher energies.

Neutron Therapy

During the late 1960s, radiotherapy with neutrons became available. At the Hammersmith Hospital, London, a neutron beam with a broad spectrum from 3–17 MeV was produced by bombarding a beryllium target with deuterons accelerated to about 30 MeV in a cyclotron.

It was initially thought that neutrons had several advantages over X-rays for treatment, in that neutron irradiation effects were independent of the oxygen concentration in the cells of a tumour. Large tumours tend to have poorer blood supplies than normal tissue and thus less oxygen, which reduces the effectiveness of X-ray treatment. Also, the quality factor (i.e. the destructive power) of neutrons is between ten and twenty times that for X-rays.

Treatments at the Hammersmith Hospital were given from a fixed horizontal beam and early results appeared promising. A second, more user-friendly, cyclotron was installed in Edinburgh which had a mobile isocentric head, but soon very severe side effects were noticed in patients.

Neutrons now appear to offer no therapeutic advantages, and are not in routine use; however, charged particle beams such as protons and pions may have clinical benefit owing to their pattern of energy deposition. (Cyclotrons are now used mainly to produce positron emitters for nuclear medicine investigations.)

Dosimetry

A major task of radiotherapy physicists is to ensure that the doses delivered to patients are as accurate as possible. This means, firstly, that radiation from the treatment machines must be measured both in terms of output and absorption in human tissue, and, secondly, that radiation treatment plans for individual patients which usually involve combinations of radiation beams, can be produced quickly and accurately.

Treatment machine output is routinely measured using small ionisation chambers with air-equivalent or tissue-equivalent walls. Such chambers are positioned in blocks of tissue-equivalent material or water (with a suitable waterproof sleeve over the chamber). Ionisation chambers used for routine machine calibration are regularly cross-checked against national standard calorimeters to ensure uniform dosimetry between treatment centres. (Ref. 11 gives an account of the procedure involved.)

Modern plastics technology has made it possible to produce so-called *phantoms* containing bony skeletons and lung substitutes for complex measurements in heterogeneous mediums.

It is always necessary to produce isodose curves for treatment planning. In the past this was achieved by laboriously moving an ionisation chamber around in a water tank to plot lines of equal dose. Since the advent of planning computers, isodose information is acquired in a more sophisticated manner. This is, typically, by measuring dose at various points along the central axis of a small range of field sizes followed by traverse plots across each beam at about five different depths. Computer controlled isodose plotters are available which can acquire data in any required configuration after initially being set up with the exploring ionisation chamber on the central axis of the beam at the position of maximum build-up dose. With linear accelerators it is usual to employ a second, reference, chamber positioned somewhere in the beam. This allows beam fluctuations to be discounted.

Computer algorithms convert these data into beam information for the complete range of available field sizes for each treatment machine. Treatment planning computer software can deal with the effects of inhomogeneities within tissue, adjusting penetration depths accordingly.

Treatment Planning

Radiotherapy treatment planning is a large subject dealt with in several text books e.g. (11), (12). The simplest form of a treatment planning for a patient involves the acquisition of an outline through the central axis of the volume to be treated and the subsequent addition of beam information.

Before the advent of planning computers in the mid 1960s, isodose curves were produced on transparencies which could be overlaid onto patient cross sectional outlines. Computers brought enormous benefits, not just by reducing the calculation time for the preparation of a plan, but by giving much more detail with great accuracy. This allowed more complex field combinations to be used providing superior treatments. Three dimensional planning became possible using outlines through other planes of the patient than the central axis; non-planar fields could be applied if necessary, and fields could be treated with different proportions of the total dose. Additionally, true allowance could be made for air gaps and inhomogeneities.These advances, in turn, encouraged radiotherapists to use ever more complex techniques, confident in the knowledge that treatment calculations would be accurate.

Figure 3.13, demonstrates a modern computer plan for the treatment of a brain tumour. The CT scan is overlaid on to the patient's outline and isodose curves are superposed and added together to complete the plan.

Recent new developments have included the use of independent beam collimators which allow the application of off-centred fields and so-called *dynamic wedges*, where the variation in position of one collimator can produce a wedged effect on the radiation beam without the addition of a physical metal wedge. Metal wedges are cumbersome and heavy for radiographers to manipulate and they introduce a degree of beam contamination.

Computer controlled multi-leaf collimators, as shown in Fig. 3.14, which can provide treatment fields of almost any required shape, revolutionised the

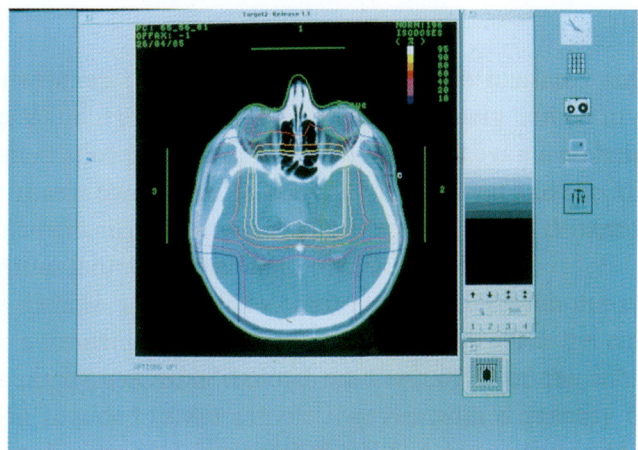

Fig. 3.13

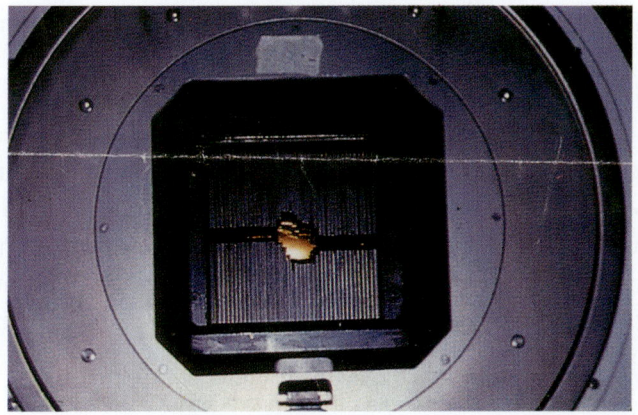

Fig. 3.14

technique of conformal therapy, previously achieved by programmed couch movements and field size alterations during treatment (13), (14).

The pilot site in the UK for the development of multi-leaf collimators in 1990 was the Christie Hospital, Manchester, where it was possible to reduce treatment volumes by up to 30% with associated reduction of the side effects of irradiating adjacent normal tissues.

Fractionation

Arguments have ensued since early in the twentieth century as to how radiotherapy should be given. The early French workers felt that many short sessions of treatment gave the best chance of irradiating cells in the sensitive phase of their cycle. The Germans favoured fewer large treatments in the belief that normal tissue had a greater capacity for repair than cancer cells.

In 1914, Henri Coutard in Paris, demonstrated that normal tissue surrounding a tumour could tolerate more radiation if it was given in short bursts, and this led to the dose fractionation schemes which have been in use since then. In Continental Europe, the United States and southern England external beam therapy is normally given in fractions of about 2 Gy per daily fraction five times per week for about 6 weeks. In Scotland, the north of England and Canada daily fractions are slightly larger, up to 3.5 Gy, five days per week for 3 or 4 weeks. The uniformity of results is explained by biological factors associated with dose per fraction, overall time and total dose. (6).

In the past 15 years, some new types of fractionation have been used. Hyperfractionation involves the use of two fractions every treatment day with reduced dose per fraction. This allows increased total tumour dose without adverse side effects. It appears to give superior results in head and neck tumours. CHART (continuous hyperfractionated accelerated radiotherapy) gives treatment three times a day for 12 days continuously and shows a significant improvement in survival rates in patients with squamous cell carcinoma.

Brachytherapy

Brachytherapy is defined as treatment using sealed radioactive sources which are placed within a tumour volume. This gives a very localised high dose and minimum dose to surrounding normal tissue.

Initial experiments with the radiations from radioactive substances, called, at the time, *becquerel rays*, were unexciting. Becquerel reported in 1898 that the germinating power of seeds could be destroyed by 'prolonged irradiation'. Pacinotti and Porcelli, in Vienna, reported a mild effect on 'various germs' irradiated with rays from uranium powder for long periods. Other workers also found that bacteria could be destroyed with high doses.

Early experiments on human skin were carried out at different centres. In January 1901, Freund, in Vienna, put small quantities of radium in packets of paper and of aluminium and bound then to his arm. After three days there was some redness which was allegedly so slight that it was thought to have been caused by the paper wrapping. However, a month later, Walkhoff, using 0.2 grams of a radium preparation on his arm for a total of 40 minutes, produced inflamed skin at 14 days, and the scar was still visible at six months. Freund repeated the experiment with painful results and hair loss. (1) pp 358–360.

Also in 1901, Becquerel himself put some radium in his waistcoat pocket and carried it round with him for a fortnight, after which he developed a severe burn on his chest. The dermatologist he consulted in Paris, Besnier, suggested that the burn was similar to an X-ray injury and that therefore radium might have a place in radiotherapy. Pierre Curie repeated the experiment on his own arm which persuaded him to provide some radium to the hospital to treat lupus erythematosis.

Malignant skin lesions were treated with radium in Vienna and in Paris in 1902. In Paris the gamma rays alone were used following work on filtration using aluminium and lead. (15). Surgeons in London and New York were initially sceptical, but radium treatment was used in the UK from about 1903.

Radium-226 is an element in the uranium decay series which starts with uranium-238 and ends with the stable isotope of lead-206. Figure 3.15, shows the decay scheme. The half life of radium is 1620 years and it decays with the emission of an alpha particle to become the inert gas, radon-222, which is also an alpha emitter. Radon too was encapsulated and used in brachytherapy

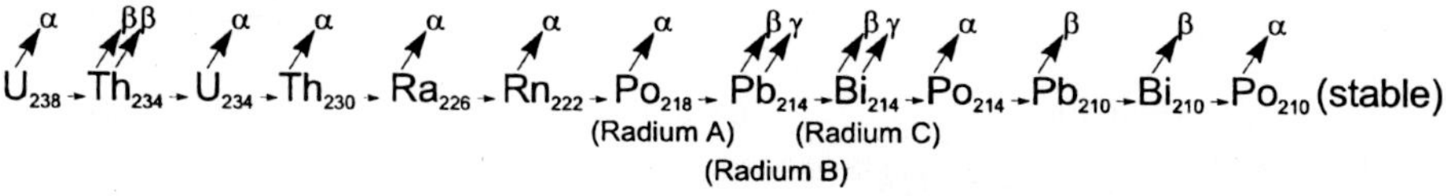

$$U_{238} \rightarrow Th_{234} \rightarrow U_{234} \rightarrow Th_{230} \rightarrow Ra_{226} \rightarrow Rn_{222} \rightarrow Po_{218} \rightarrow Pb_{214} \rightarrow Bi_{214} \rightarrow Po_{214} \rightarrow Pb_{210} \rightarrow Bi_{210} \rightarrow Po_{210} \text{(stable)}$$

(Radium A)   (Radium C)

(Radium B)

Fig. 3.15

regularly until the 1970s. Since neither of these elements emits gamma rays, which alone have the penetrating power for this type of treatment, it might be asked why they were used so successfully. The answer is found in their decay products, lead-214 and bismuth-214, (sometimes called radium B and radium C respectively), neither of which can be used on its own because of very rapid decay, but both of which emit energetic gamma rays. They also emit unwanted beta particles, but the metal casing, typically a platinum-iridium alloy, serves the dual purpose of providing a strong and gas tight cover and an absorber of soft betas.

Radium was a very practical brachytherapy source. It was available in reasonable quantities and with high enough activity per unit mass, (specific activity) for successful treatment. However, it does have disadvantages. Its maximum gamma ray energy of 2.4 MeV means that the half value layer in lead is 16 mm, giving problems with the provision of shielding for staff. Lead shields, even when fitted with wheels, are heavy for nurses and others to move. Radon gas can also be a hazard if radium needles or tubes are damaged.

The advent of artificially produced radionuclides from nuclear reactors and cyclotrons had a profound effect on brachytherapy. Many of the new isotopes produced had more suitable characteristics for treatment. The use of radium diminished steadily in the 1960s and has since been strongly discouraged.

The most commonly used brachytherapy sources at present are as follows:

| Radionuclide | Photon Energy (MeV) | | Half life |
	Mean	Max	
Cobalt-60	1.25	1.33	5.27 years
Caesium-137	0.662	0.662	30.0 years
Iridium-192	0.37	0.61	74 days
Iodine-125	0.028	0.035	60 days
Gold-198	0.42	0.68	2.7 days
compared to Radium-226		2.40	1600 years

Cobalt-60 is not now often used for brachytherapy although it has been used in needles and tubes for gynaecological treatment and in ophthalmic applicators. At one time cobalt wire was available, but it was discontinued because it was brittle.

The first remotely controlled machine for treating cancer of the cervix, the Cathetron, employed cobalt-60 pellets of very high specific activity. It was developed by O'Connell and Joslin at Charing Cross Hospital, London, and was commissioned in 1966. Previous treatments were based on radium with dose rates only a small fraction of this, and some radiotherapists attributed the good results from radium therapy to the low dose rate regime. Typically, three insertions of about 24 hours each were made over three weeks, whereas with the Cathetron the insertion time was measured in minutes. Clinical results were, however, as good as those from low dose rate treatments, although the arguments about relative effectiveness still go on.

Caesium-137 is the most commonly used source today. It comes in the form of needles for implants, or tubes for intracavitary gynaecological treatment. Most gynaecological treatments nowadays are provided via automatic afterloading of caesium sources from purpose built systems. In these units the caesium is in the form of spherical balls, and both low- and high-dose-rate machines are available. The half life of 30 years means that sources can be used for several years, but the normal working life is recommended by the manufacturers to be about ten years because of wear to the encapsulation.

Iridium-192 is usually provided in the form of wire hairpins or single strands encapsulated in 0.1 mm of platinum which absorbs the beta particles. The wire can be cut to length for insertion. It is often used for implanting into the breast. Pellets of high specific activity are also available for automatic afterloading machines in some treatment centres.

Tantalum-182 was the first radionuclide to be fashioned into hairpins in the 1950s. With a half life of 115 days and a maximum photon energy of 1.29 MeV it was used with success to treat tumours requiring flexible sources. It went out of favour when iridium became available more cheaply.

Iodine-125 decays by electron capture producing characteristic X-radiation with principle energies of 27 and 31 keV and 35.5 keV gammas. The iodine is either in ion exchange resin balls or adsorbed onto silver and then encased in titanium. The resulting seeds are used for temporary or permanent implants.

Small gold-198 grains or seeds are encapsulated in platinum and are usually used for permanent implants because of the very short half life. They have also been used for small area treatments in moulds.

References

1. *Elements of General Radiotherapy for Practitioners* L. Freund (translated by Lancashire) Rebman Ltd., London, 1904.
2. Despeignes. *Lyon Med. J.* Dec. 1896.
3. Johnson W. and Merrill W. H., *Phil Med. J.* **6**, 1089–1091, 1900.
4. Dische S. and Saunders M. I., Clinical Fractionation Studies. *Oxford Textbook of Oncology*, ed. Peckham *et al.*, Oxford University Press. Oxford, 1995.
5. Franklin M., Method of Measuring X-rays. *New York Med. J.* **81**, 802–807, 1905.
6. Perry J. A., *RPL Dosimetry,* Bristol, IOP Publishing, 1987, ISBN 0-85274-272-X.
7. McKeever S. W. S., Moscovitch M. and Townsend P. D., *Thermoluminescense Dosimetry and Materials*, Ashford, Nuclear Technology Publishing, 1995, ISBN 1-870965-19-1.
8. Van de Graaff R. J., A 1,500,000 volt electrostatic generator. *Phys. Rev.* **38**, 1919–1920, 1931.
9. Van de Graaff R. J., The electrostatic production of high voltage for nuclear investigations. *Phys. Rev.* **43**, 1933.
10. *Physics of Electron Beam Therapy,* S. C. Klevenhagen, Adam Hilger Bristol, 1985.
11. *Radiotherapy Physics,* eds. J. R. Williams and D. I. Thwaites, Oxford Medical Publications, 1993, ISBN 0-19-963315-0.
12. *Radiotherapy Treatment Planning*, R. F. Mould, Adam Hilger, Bristol, 1990, ISBN 0-95274-504-4.

13. Cotter G., Adjustable field shaping for external beam radiation therapy. *Radiology* **174(3)**, 892–893, 1990.
14. Yu C. X. *et al.*, A method for implementing dynamic photon beam intensity modulation using independent jaws and a multi-leaf collimator. *P. M. B.* **40(5)**, 769–787, 1995.
15. Radium Rays for Cancer. *Medical Record*, 63, 64, 1903.

Chapter 4

TREATMENT WITH NON-IONISING RADIATION

Figure 4.1, is a diagram of the electromagnetic spectrum. Chapter 3 has dealt with treatment with radiations of wavelengths up to about 100 nm which cause ionisation in tissues. Longer wavelengths produce different effects in the body.

This chapter will deal with non-ionising radiations in order of increasing wavelength.

Molecular Effects of Non-Ionising Radiation

When non-ionising radiation is absorbed it is converted into vibrational and rotational energy, and there is a change in the electronic configuration of molecules. In its ground state, internal motion within a molecule is minimal. Absorption of non-ionising radiation converts the molecule to an excited state by raising an orbital electron from the ground state (s_0) to the first excited singlet state (s_1), and any remaining energy increases the vibrational state of the molecule. Some heat may be generated. Return to the ground state (de-excitation) can occur by molecular collision or by the emission of radiation of a greater wavelength, resulting in fluorescence.

Some molecules are capable of radiating energy from a state between the ground state and excited singlet state. When this happens the electron spin is reversed so that the molecule contains two electrons with unpaired spins.

The effect of non-ionising radiation on a cell depends upon the chemical composition of the cell. Radiation must be absorbed before it can cause a change, and molecules in excited electronic states have different chemical and physical properties to those in ground states. Depending upon its

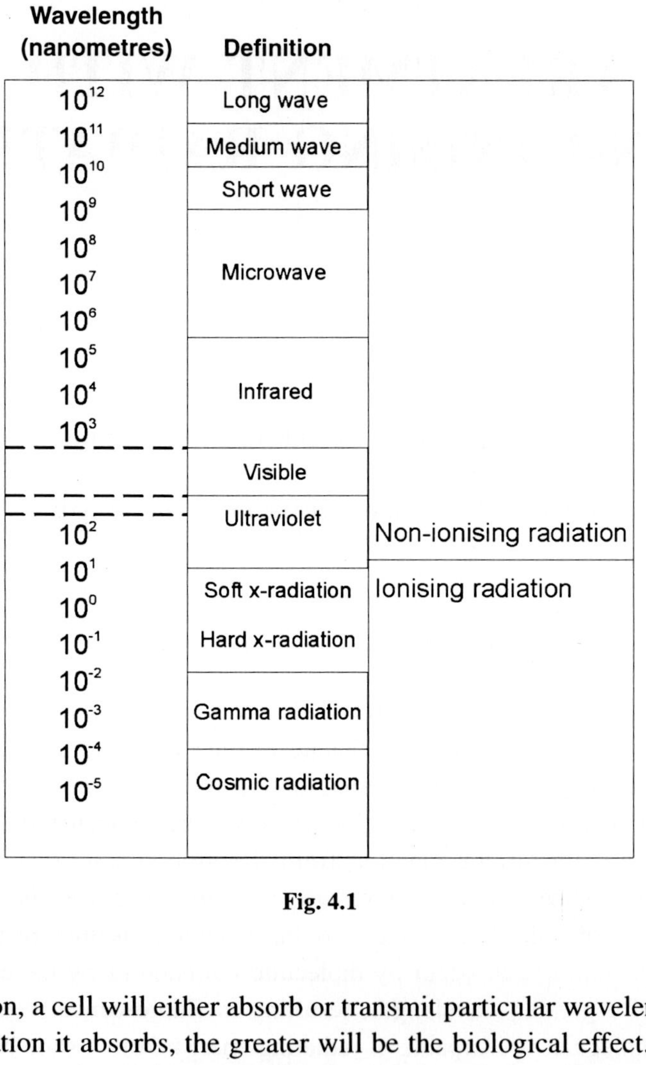

Wavelength
(nanometres)   Definition

Wavelength (nanometres)	Definition	
10^{12}	Long wave	
10^{11}	Medium wave	
10^{10}	Short wave	
10^{9}		
10^{8}	Microwave	
10^{7}		
10^{6}		
10^{5}		
10^{4}	Infrared	
10^{3}		
	Visible	
10^{2}	Ultraviolet	Non-ionising radiation
10^{1}		
10^{0}	Soft x-radiation	Ionising radiation
10^{-1}	Hard x-radiation	
10^{-2}		
10^{-3}	Gamma radiation	
10^{-4}		
10^{-5}	Cosmic radiation	

Fig. 4.1

composition, a cell will either absorb or transmit particular wavelengths. The more radiation it absorbs, the greater will be the biological effect. The term *action spectrum* is used as an indication of the relative effect of different wavelengths on a particular chemical compound. For more detail on the effects of radiation on tissues the reader is referred to Refs. (1) and (2).

The penetration depth in the skin of various wavelengths is shown in Fig. 4.2. Wavelengths between 600 and 1400 nm can penetrate beyond the dermis.

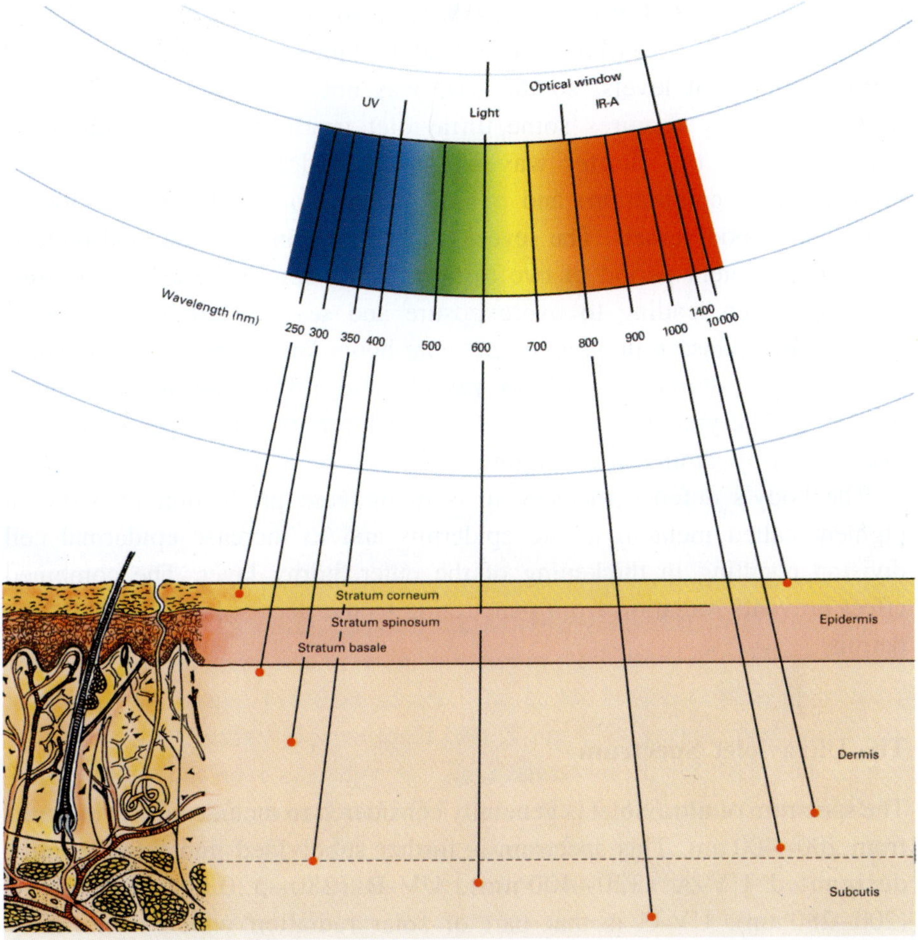

Fig. 4.2. Skin penetration depth of optical radiation.

Ultraviolet Radiation

The chief beneficial effect of ultraviolet radiation on the body is the production of vitamin D in the skin. Ultraviolet irradiation brings about the formation of calciferol (vitamin D_3) in the layer of malpighian cells in the lower epidermis. This vitamin is important for calcium absorption in the body and only a small proportion of the amount required is supplied by the diet. Vitamin D-fortified foods have been tried in the United States but, unfortunately, overdosage can lead to kidney damage and elevated serum cholesterol levels, so the trial was not successful, but although the human body requires some ultraviolet irradiation, too much of it, particularly by the shorter wavelengths, has deleterious effects. Skin erythema (red colouration) and painful oedema occur due to dilation of superficial blood vessels. The severity depends upon the dose, and there is an insidious latent period of two to three hours before the effects become apparent, often leading to overexposure and severe sunburn. A second erythemal response following six or more hours later is more prolonged. It is caused by damage to blood vessels and inflammation. At longer wavelengths, erythema takes longer to develop but is more persistent than that caused by shorter wavelengths.

The body's defence mechanism is to increase production of a brown pigment called melanin in the epidermis and to increase epidermal cell division resulting in thickening of the outer, horny layer. The combined effect prevents radiation from penetrating to the deeper layers of skin, the dermis.

The Ultraviolet Spectrum

The spectrum of ultraviolet is generally considered to include all wavelengths from 200–400 nm. This spectrum is further subdivided into three regions, designated UV-A (320–400 nm), UV-B (280–320 nm) and UV-C (200–280 nm). UV-C is that part of solar radiation which is normally filtered out by the ozone layer in the atmosphere before it can reach the earth's surface. UV-C is also called germicidal radiation because it kills small organisms. The remaining two regions are so divided because the effect on human skin of UV-B is several orders of magnitude more damaging than UV-A. UV-B is most often responsible for sunburn, however, UV-A is thought to be responsible for longer term effects such as skin ageing and loss of elasticity.

The beneficial and harmful effects of sunlight have been known for many centuries and sunlight has been used for treatment throughout history.

Phototherapy and Photochemotherapy

Phototherapy means treatment using non-ionising radiation. Photo-chemotherapy means treatment with a combination of chemical substances and radiation.

In ancient Egypt around 1500 BC it was found that some skin problems could be improved by exposure to sunlight. It was also established that the effect could sometimes be accelerated by eating certain plant seeds and extracts before exposure.

The ancient Greeks were in the habit of sunning themselves on the flat roofs of their houses, having first covered themselves with oil. Hippocrates established a clinic on the island of Cos in about 420 BC which included a sanatorium with an open gallery facing south for patients to receive sunlight treatment. This was often for pleasure but also to improve their health. The Romans also took sun baths, constructing special outbuildings, or *solaria*, where this treatment, called *heliosis*, could take place. Herodotus mentioned the use of sun baths for people with 'enfeebled muscles', while Antyllus described erythema and suggested the use of sun baths for 'loss of fat, lessening of swellings, improvement in cases of dropsy, sciatica, affections of the kidney, chronic diseases of the bladder, and paralysis'.

Heliotherapy is the term now used for phototherapy using sunlight as the source of radiation.

Not until the early nineteenth century are there any records of *in vitro* work. In the 1880s, bacteriologists proved that light could kill various 'germs', and they found that rays at the blue end of the spectrum were the most effective. From this date sun baths were given for obesity, diabetes, gout, and rheumatic disorders, as well as for rickets (caused by lack of vitamin D).

The leading authority on phototherapy in the late nineteenth century was Niels Finsen, a Danish physician. His experiments on his own arms enabled him to establish conclusively that ultraviolet irradiation produced erythema. He covered various skin areas with coloured glass filters and with rock crystal (quartz) which allows transmission of ultraviolet radiation, and measured the effects of the different wavelengths. Finsen found that the protective effect of dark pigment could be replicated by the application of dark paint. He, and others, also experimented on penetration depths of a range of wavelengths using human skin freshly cut from burn blisters and the blisters of patients with pemphigus vulgaris. This disease is characterised by blisters whose surface can be of varying thickness. (3).

Finsen is best remembered for his successful treatment of lupus vulgaris (tuberculosis of the skin), and was awarded the 1903 Nobel Prize for Medicine in recognition of this work. His early treatments used sunlight which was focused through a quartz lens cooled with water. As a more constant radiation source, he developed a carbon arc lamp for ultraviolet treatment. This was described in a 1904 brochure (4), thus:-

'The Finsen Arc-Lamp consists of an arc lamp of 60 to 80 amperes in which the positive carbon is placed above the negative, so that the rays, in passing from the crater, are projected downward and outward. In the axis of the rays issuing from the luminous point are four copper telescopic tubes the lenses of which are made of rock-crystal so as not to intercept the chemical rays (ultraviolet), each one provided with water circulation for cooling the rays as they pass out.'

Treatment of four patients simultaneously was possible, and Fig. 4.3 shows the equipment used by Finsen in his clinic in Copenhagen.

A rather more portable carbon arc irradiator, produced in London by Watson, is shown in Fig. 4.4. This ran at 20 amps and the carbon electrodes were angled towards each other. Radiation was directed at the patient via a quartz lens. A larger area of skin could be treated than with the Finsen lamp.

Carbon arc lamps were not popular with their operators because they emitted noise, sparks and odour. They were superseded by mercury arc lamps like that shown in Fig. 4.5 being used as a treatment for ringworm. The spectrum emitted from mercury lamps depends upon the pressure within

Fig. 4.3. Finsen's apparatus.

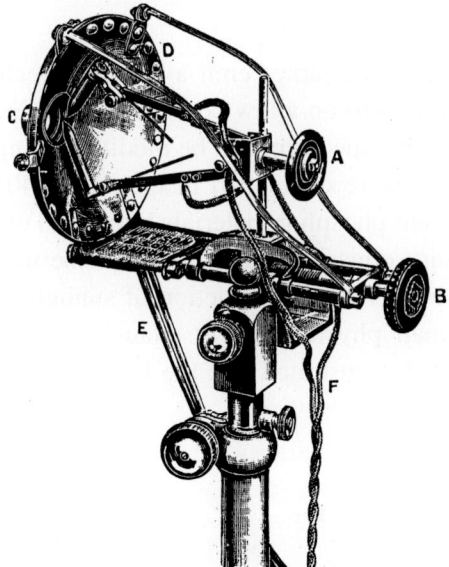

Fig. 4.4. The apparatus of Lortet & Genoud. (*Watson & Sons*, London).

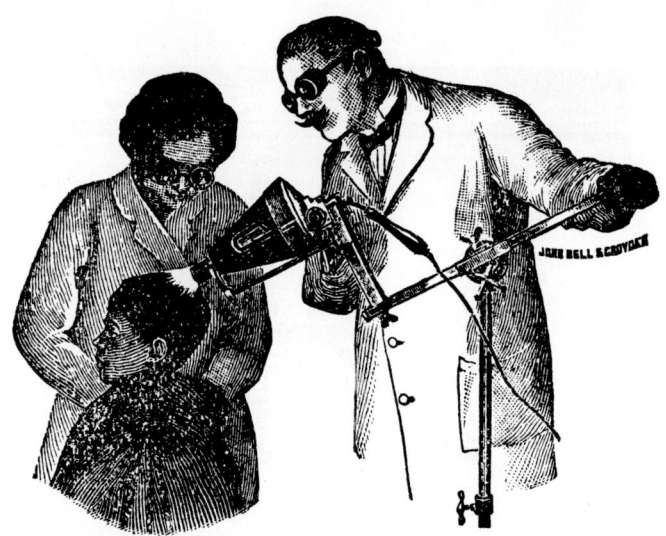

" ARNOLD " TUNGSTEN ARC LAMP IN USE.

Fig. 4.5

the lamp. Medium pressure lamps emit a band of spectral lines from about 240–400 nm superimposed on a low continuum, thus including UV-A, UV-B, and some UV-C. The spectrum can be modified using appropriate filters.

Fluorescent lamps were developed in the late 1940s and, since then, a variety of envelope and phosphor materials have been used to produce lamps with all the necessary emission wavelengths in the ultraviolet region.

The action spectrum for the production of sunburn was first established in 1921 by a German physicist, K. W. Hauser, who had built himself a monochromator and demonstrated that the UV-B wavelengths were chiefly responsible.

Medical Uses of Ultraviolet

Ultraviolet irradiation was found to be useful in alleviating a whole range of skin conditions. Early workers treated roseola (a childhood viral infection producing a high fever and giving rise to an unpleasant pink pustular rash), and the facial scars of patients who had survived smallpox, in addition to lupus vulgaris. Over time, more skin conditions were treated including acne, excema, pruritis, ringworm, and varicose ulcers. The two most commonly treated conditions are psoriasis and vitiligo.

Psoriasis is an extremely common and distressing disease in which the cell cycle is greatly accelerated in the skin. From their creation in the dermis to the time they are shed as dead skin by the epidermis, normal skin cells take about 28 days. In psoriasis this time is decreased to as little as 3–4 days. The effect is that sufferers develop unsightly and uncomfortable scaly lesions (sometimes called plaques), often over large areas of the body, and flakes of dead skin are shed continually.

In vitiligo, patients have patches of unpigmented skin which is unsightly. The unpigmented areas burn and blister rapidly on exposure to the sun, but it has long been known that after recovery, the skin develops some pigmentation.

In 1923, H. E. Alderson, (5) described the use of a quartz filtered mercury discharge lamp to treat psoriasis. A few years later, phototherapy was used in conjunction with the application of coal tar preparations which made the treatment more effective. It was many years before the most effective wavelengths for clearing the skin were discovered. In 1966, Bowers and colleagues at Gloucester, England, found that the band from 290–320 nm was effective whereas longer wavelengths were ineffective. They also discovered that UV-B plus tar or certain drugs was more effective than irradiation alone. Swedish workers used 313 nm radiation from a mercury lamp effectively, while Parrish and colleagues in America noted that UV-A alone could be used but in very large doses.

In the 1940s a Cairo dermatologist, El Mofty, became aware that patients suffering from vitiligo could increase the effectiveness of sunburn treatment by first eating the seeds of a plant called *ammi majus* which grew beside the river Nile. El Mofty analysed the seeds and isolated the three-ringed organic chemical compounds which became called psoralens (6). These compounds were also found in other plants, notably, celery, lemon, lime, parsnip, and parsley, and in greater abundance in bergamot and clove.

PUVA Treatment

Since UV-A was considered to be the least damaging part of the UV spectrum, the combination of UV-A plus psoralen was introduced as a treatment for both conditions mentioned above and for a number of other skin diseases including polymorphic light eruption and some low grade malignancies like mycosis fungoides. It is also found to be effective for restoring hair growth in some patients.

The mechanisms of UV-A and PUVA effects on tissues are poorly understood. Ultraviolet irradiation causes molecular excitation, generating free radicals leading to molecular damage. This can lead to distortion of DNA strands and interference with DNA processes. The effect is to slow down cell reproduction which is exactly what is needed in the treatment of psoriasis and some other conditions.

Psoralens are relatively rigid 3-ringed chemical compounds which allow the formation of biochemical links. Ultraviolet energy weakens the structure making psoralens chemically active. Psoralen molecules then bind to DNA in living cells and cause further slowing of the cell cycle.

It is important to note that although UV and PUVA treatments clear the skin, they do not cure the disease. The skin of psoriasis patients can normally be cleared by twice weekly treatments for about six weeks. Thereafter it is usually necessary for patients to undertake regular, but less frequent, sessions to maintain this position.

Originally, psoralens were administered orally. A dose of 8-methoxypsoralen (8-MOP), depending upon body weight or surface area, was given one and a half or two hours before exposure to UV-A. This is still the treatment of choice for a large number of clinical situations but it does have some disadvantages. Patients may suffer side effects such as nausea or pruritis, and hepatic damage is possible. Following ingestion, the total skin surface and the eyes of patients are particularly sensitive to normal sunlight. Cataracts can be induced in patients who do not wear appropriate eye protection for about 24 hours after taking the drug.

Recent successful attempts to avoid some of these side effects include bathing patients in water containing psoralens before treatment, and the topical application of psoralens in oils or gels. Bath PUVA has resulted in lower subsequent radiation doses for the same cosmetic effect, and oils or gels are particularly useful when only small areas require treatment.

There has always been concern that exposure to ultraviolet wavelengths can cause cancer. Ultraviolet has been implicated in all types of skin cancer, of which malignant melanoma is the most serious. High cumulative doses are also associated with other skin problems such as loss of elasticity. Treatment with UV-A and UV-B before about 1970 was an empirical science. Exposure was usually timed but little account was taken of actual radiation dose. Fortunately, physicists were invited to undertake measurements once PUVA treatment became widespread, and there is now clear evidence that cumulative dose above about 1000 Joules per square centimetre results in a risk of pre-malignant or malignant skin changes. Accurate dosimetry is very important and regular machine calibrations and uniformity checks must be carried out if treatment is to be given responsibly. Not only does this ensure that patients receive the correct prescribed dose so allowing treatment regimes to be optimally effective, but also to maintain accurate records of patients' lifetime exposure to ultraviolet radiation.

Narrow Band UV-B

Patients requiring long term treatment inevitably amass high cumulative doses of UV radiation. It was apparent from early days that the most effective wavelengths for the production of erythema and sunburn were in the UV-B range. Serious attempts were made from the 1980s to determine which actual wavelengths of UV actually contributed to treatment and which contributed to unwanted effects.

As a result of this research, van Weelden and colleagues in the Netherlands (7), developed a very narrow band UV-B tube centred around 311–313 nm and demonstrated that this was as effective as PUVA in the treatment of psoriasis. It is more convenient for patients since no drug is involved, and it is thought to be less carcinogenic. Narrow band UV-B is now widely available as an alternative to PUVA.

The Physics of Quality Control

For accurate treatment:

1. Spectral output must be consistent.
2. Radiation dose must be accurate.
3. The radiation must be uniform over the treatment area.

The Measurement of Wavelength

Spectral distribution is measured with a spectroradiometer for which the optics are shown schematically in Fig. 4.6. The radiation enters through a narrow slit into a light tight box where it is focused on to a diffraction grating. The angle of incidence on the grating determines the wavelength or wavelengths which are diffracted to the output optics. The grating is ruled with a very large number of parallel grooves like those shown in Fig. 4.7(a). Consider light rays A and B of wavelength w incident on adjacent grooves.

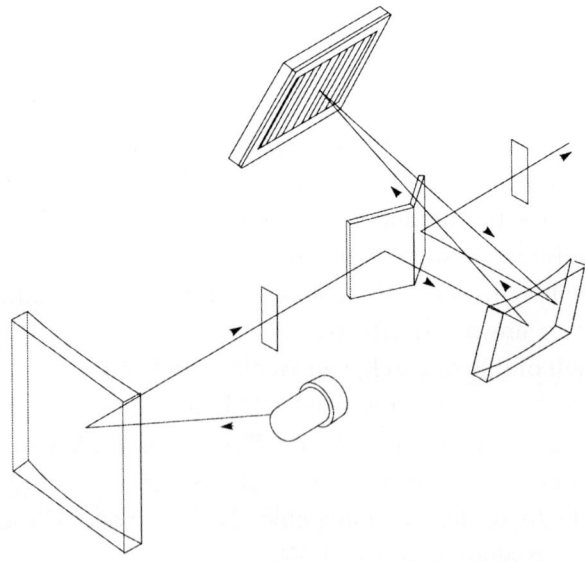

Fig. 4.6

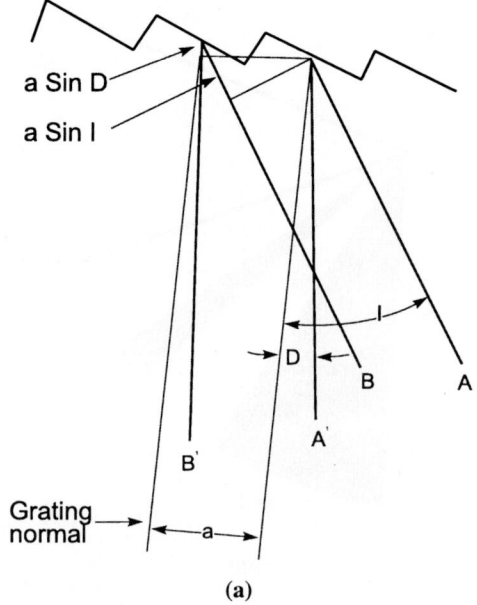

a Sin D

a Sin I

D

B'

A'

B

A

I

Grating
normal

a

(a)

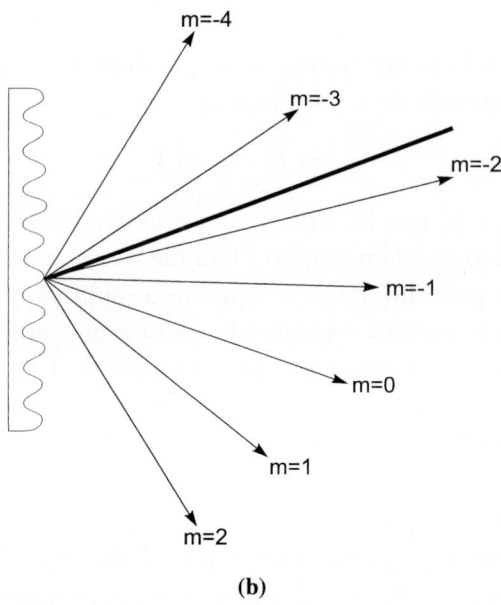

m=-4

m=-3

m=-2

m=-1

m=0

m=1

m=2

(b)

Fig. 4.7

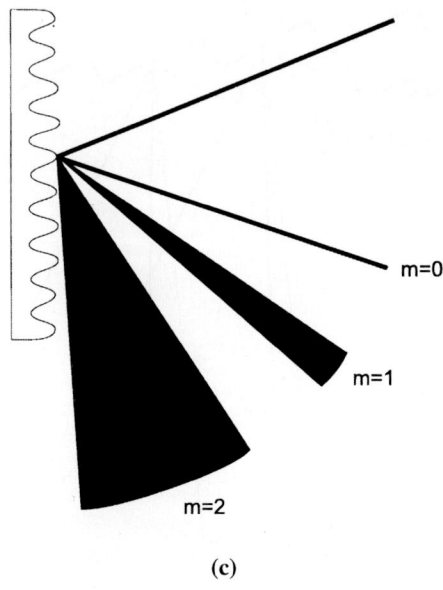

(c)

Fig. 4.7 (*Continued*)

Some light at angle D to the grating normal is shown as rays A^I and B^I. The path difference between these two rays is

$$a \sin D + a \sin I$$

Summation of rays A^I and B^I will give destructive interference if the path difference is equal to an odd multiple of half the wavelength, and constructive interference if the path difference is equal to a multiple of the wavelength.

Figure 4.7(b), shows how a parallel beam of monochromatic light is diffracted in directions corresponding to multiples 0, 1, 2, etc., where the integers are referred to as the order of diffraction. When a parallel beam of polychromatic light is incident on the grating, the light is dispersed as shown in Fig. 4.7(c).

It will be apparent that light of diffraction order 1 for one wavelength may have the same path as light of order n for another wavelength. For more complete separation of the wavelengths in a spectroradiometer, a double monochromator is required, so that diffracted light passes through another

slit into a second monochromator so removing second and third order spectra
before the wavelength required is extracted through a final exit slit.

The spectral output of the two types of lamps used for PUVA therapy
varies little and is very predictable. Fluorescent tube irradiators have a coating
of either barium silicate or strontium borate on the inside of the glass. This
is easy to apply consistently during manufacture. Mercury discharge lamps
have the mercury line spectrum superimposed on a continuum which varies
only if the pressure in the lamp changes. Figure 4.8, shows the spectra of
these two types of irradiator.

Routine Dose Measurement

Almost any detector which is used to measure visible light, can also be used
with ultraviolet radiation. Typical detectors which have been used include

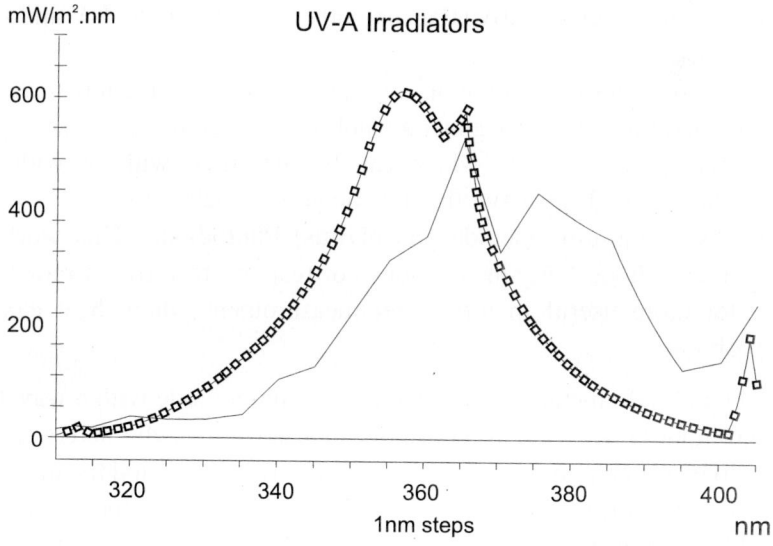

Spectral output from: fluorescent UV-A tube (boxes) and quartz
halogen lamp (line).

Fig. 4.8

thermopiles, bolometers, pyroelectric detectors, photodiodes, photomultipliers and UV sensitive films. Thermopiles and bolometers are absolute detectors which are linear in wavelength response and have a linear relationship with intensity. They are, however, expensive and relatively fragile and can be difficult to use.

Routine measurements can be made using calibrated photon detectors. Examples of photon detectors and their mode of operation are:

1. Vacuum Photodiode. Relatively energetic photons (from UV to near infrared) cause emission of electrons when incident upon surfaces in a vacuum. The electron yield is high when particularly sensitive materials are used as the target. An applied voltage (bias) causes electrons to flow and creates a current which is proportional to light intensity. Vacuum photodiodes are useful at high signal levels but may be swamped by noise at low light levels.

2. Photomultiplier. This is a combination of a vacuum photodiode and an electron multiplier. The resulting detector is at least two orders of magnitude more sensitive than any other detector in the UV and visible regions.

3. Silicon Photodiode. Photons impinge on a p-n junction in a semi-conductor material to generate holes and electrons. These create a current and a voltage which can be measured with or without an applied bias. They have the advantage of small size.

4. GaAsP (Gallium arsenide phosphorus) Photodiode. This works like a silicon diode but, since it does not respond to infrared radiation, is a lot more useful in ultraviolet measurement where heat can be a problem.

A broad band radiometer combines a detector photodiode with a wavelength filter and input optics such as a cosine response filter. Radiometers are often claimed to be suitable for UV-A or for UV-B or for UV-C, and the implication is therefore that a UV-A radiometer would have a uniform spectral response from 320–400 nm with no response outside these wavelengths. In practice, all radiometers of this type have non-uniform spectral sensitivity. This is why broad band radiometers must be calibrated for the type of source spectrum with which they are used.

Calibration is achieved by using a calibrated spectroradiometer to determine the total irradiance within the waveband of interest, and setting the radiometer display accordingly.

Safety

Staff should avoid personal exposure where possible, and need to wear appropriate clothing and eye protection when working near energised irradiators. Some cosmetics and perfumes contain photosensitisers which should be avoided. Photochemotherapy patients need protection from sunlight while photosensitisers are active. Eyewear is tested for effectiveness using a UV source and a radiometer.

UV lamps contain gas at high pressure giving a risk of explosion if accidentally knocked.

For more detailed study the reader is referred to Refs. (8)–(10). Full information on the measurement of ultraviolet radiation can be found in (11).

Visible Light

The majority of medical physics involvement with visible light concerns lasers, but there are other therapeutic applications.

Blue Light Phototherapy

Neonatal hyperbilirubinaemia (jaundice) is a common condition in premature and very small babies. When red blood cells are destroyed, the body scavenges iron from them producing bilirubin. This is useful in small quantities because it is a strong antioxidant although it is toxic. Normally the liver processes bilirubin and excretes it in bile where it is broken down by bacteria in the gut. Small infants have immature livers which are incapable of conjugating bilirubin, so serum bilirubin increases producing jaundice which can lead to brain damage if it persists.

Blue light, from 400–500 nm, has three different effects on bilirubin; photo-oxidation, configurational photo-isomerisation (which makes bilirubin less toxic) and structural photo-isomerisation (which changes the structure of pigment molecules). These processes make bilirubin more soluble in water and thus less difficult to excrete.

Blue light is commonly administered from fluorescent tubes or metal halide lamps and it is important to ensure that the spectrum contains no ultraviolet components.

An inherited condition called Crigler-Najjar Syndrome results in a deficiency of a particular hepatic enzyme. This syndrome can be controlled by regular blue light treatment, but banks of tubes are required for larger patients.

Seasonal Affective Disorder (S.A.D.)

The application of high intensity, broad-band white light (> 2500 Lux) is an effective treatment for winter depression which is thought to be caused by an excess of the hormone, melatonin. Bright light suppresses melatonin production.

This form of light therapy has also been used to treat severe jet-lag and 'shift-worker syndrome'.

Lasers

The laser was developed as a result of an entirely new concept in the production of electromagnetic radiation. Instead of using the energy of free electrons moving among atoms, the new invention utilised energy states within atoms to produce electromagnetic waves.

The first device capable of operating on this principle involved microwaves with wavelengths of the order of centimetres. This was called a MASER — an acronym for Microwave Amplification by Stimulated Emission of Radiation. It was developed by Professor Charles H. Townes and colleagues at Columbia University in the United States (12).

Einstein had proposed the Quantum Theory of radiation in 1917 suggesting that energy could be added to raise an atom from its ground state to an excited state only in certain well defined amounts. A photon of exactly the right frequency, and therefore the appropriate energy to bridge the gap between two energy levels, would raise the atom to an excited state. Einstein postulated that an atom in an excited state may be stimulated to emit a photon in response to the incoming photon.

Normally, most of the atoms and molecules in a system are in lower energy states, so that a photon of the proper frequency entering is likely to be absorbed. Even if it were to strike an excited system and produce two free photons, these would be absorbed because there would be more atoms or molecules in unexcited states. However, if photons of the proper frequency enter a medium in which most of the atoms are already excited, more photons will be released than those entering.

Since all electromagnetic radiation consists of a stream of photons, the principle leads to a new method of amplifying electromagnetic radiation.

Historical Development

Townes and his colleagues were studying the absorption of microwaves in gases including ammonia. The ammonia molecule, made up of one nitrogen and three hydrogen atoms, has two energy states. At the higher energy level, the molecule is repelled, but at the lower level attracted by a strong electrostatic field. The difference in energy levels is equivalent to the energy of a photon of frequency 24,000 megahertz.

He separated the molecules by forcing a stream of ammonia through a cylinder made from charged rods, where the electrostatic field was high close to the rods but weak near the centre of the cylinder. By bombarding the upper level molecules with microwaves at 24,000 megahertz, they gave up their energy in the form of additional photons at the same frequency, thus amplifying the incoming wave and producing a coherent beam of monochromatic energy.

The excited molecules were directed into a highly reflective chamber in which the photons were reflected many times interacting with other molecules and producing yet more photons.

The gas maser was followed by the development of a solid-state maser, which was more suitable as an amplifier. It had a very low noise level and a greater power output.

In 1958, a paper by A. L. Schawlow and C. H. Townes (13) proposed a method of constructing a maser to operate in the visible and infrared regions of the spectrum. This would use a resonant cavity with optically reflective surfaces at both ends, but with one of the mirrors partially transparent to allow the radiation beam to emerge. Intensive research programmes were organised in universities and industrial laboratories resulting, in 1960, in the publication of the first paper on the successful development of an optical maser by T. H. Maiman of the Hughes Aircraft Laboratories (14). This produced a deep red, 694 nm, pulsed beam using a cylinder of pink ruby as the excited medium. The term, LASER, replacing the 'm' for microwaves with an 'l' for light was not used until a year or so later. Townes and two colleagues received the 1964 Nobel Prize for physics.

The first gas laser was invented at the Bell Telephone Laboratories a few months later by A. Javan. This consisted of a plasma tube containing a mixture of helium and neon, in the ratio 10:1, and it produced a coherent beam of red light at 633 nm. Work with other inert gases allowed William Bridges to develop the argon laser in 1964 (15) giving two lines at 488 and 514 nm in the blue and blue-green regions of the spectrum.

The principle components of a laser are shown diagrammatically in Fig. 4.9.

Clinical Development

The earliest clinical users of ruby lasers were ophthalmologists who experimented on animals and produced retinal burns. This led to the successful treatment of retinal detachment, a very common and potentially blinding

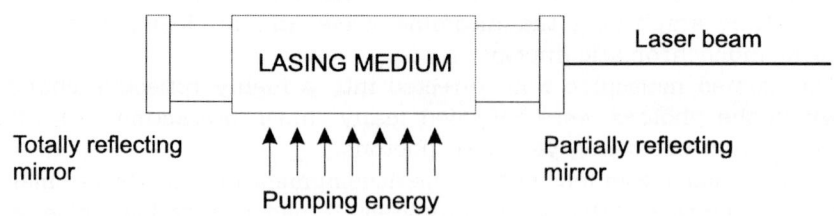

Components of a laser

Fig. 4.9

condition, in humans. Holes and tears in the retina could previously be treated by open surgery, or the tear could be "welded" with diathermy or a xenon flash tube. The ruby laser acted in the same way as the xenon flash tube, but could repair the condition in about a millisecond instead of more than a second.

The ruby laser was used in 1963 at the New England Medical Center on various malignant tumours in animals with success. Human trials began with the removal of skin cancers in 1965, but the technique received a setback when it was found that viable cancer cells were ejected explosively during the pulsed irradiation of skin tumours. These cells had the potential of re-implanting in surrounding normal tissue, and there were also fears for the safety of staff involved. It appeared that laser surgery would not be suitable for the eradication of malignancies.

The CO_2 laser, invented by another researcher at Bell Telephone Laboratories, C. K. N. Patel, in 1965 (16), was not a pulsed laser but one with continuous operation and an output at 10,600 nm in the far infra-red. This wavelength is heavily absorbed by water so that the depth of cut in tissue can be controlled. The laser was found to be very efficient and extremely powerful with an output of many watts — several orders of magnitude higher than the pulsed ruby laser. Additionally, extensive tests found no viable cells in the debris emitted during surgery, and the CO_2 laser was introduced clinically in 1970. Its chief disadvantage is the total absorption in glass making the use of fibre optic transport impossible; a system of mirrors in an articulated arm is required for use in surgery.

Continuous Wave and Pulsed Lasers

A laser may have continuous or pulsed output depending upon its method of operation. The exposure time of a continuous wave (CW) laser is determined by a shutter which controls the exposure, in milliseconds, electronically. Lasers which are pumped by a flashlamp give pulses whose duration depends upon the characteristics of that lamp. Lasers which have an R-F power source have a rapidly oscillating output but their function is similar to CW lasers.

Nanosecond pulses can be produced using the technique of Q-switching. The name derives from a concept of quality factor because less energy is wasted. The resonance characteristics of the lasing cavity are altered using an electro-optic or acousto-optic shutter, and, because laser energy is emitted in short time periods only, the peak power can be very high, in the region of megawatts or gigawatts.

Effects of Lasers

Thermal Effects

As the temperature of the irradiated tissue is raised by absorption of laser energy, it passes through the various stages of denaturation, coagulation, vacuolation, vaporisation, carbonisation and incandescence. Coagulation and vaporisation are the most significant, and cutting tissue with a laser is a form of vaporisation.

Water boils at 100°C, but other tissue components vaporise at higher temperatures. When water has been evaporated off, carbon may be left behind, and carbon particles absorb energy well, vaporising at an extremely high temperature. The heat generated is dissipated not only in air but also into surrounding tissues producing secondary thermal effects.

Coagulation produced on the way to vaporisation leads to haemostasis and the prevention of bleeding, and the rate of delivery of laser energy causes variation in the ultimate effect.

Non-Thermal Effects

The chief non-thermal effects of laser irradiation are the photo-mechanical effect and the photo-chemical effect. The first is limited to high power beams of very short pulses such as the Q-switched laser. As the laser pulse impinges on tissue, the energy, through thermal expansion, produces waves (acoustic transients) which can rip tissue apart. If this happens near the surface, plumes of smoke and debris can be ejected.

Photo-chemical effects have already been mentioned in connection with ultra-violet treatment. These effects result in the activation of molecules in tissue by chemical stimulation, and can be equally useful in laser treatments. The technique is called photodynamic therapy (PDT).

Applications

There are now many types of laser offering a complete range of wavelengths for different surgical uses. The active medium may be solid, liquid, gas or semiconductor, and the effect on tissue depends upon the amount of water, haemoglobin, calcium, pigment, fat, and so on, present.

Figure 4.10 shows how various lasers used in clinical practice are positioned in the spectrum. Overlaid on the spectrum are the absorption curves for haemoglobin, melanin and water which demonstrate why different wavelengths have differing clinical effects on tissue. The two peaks for absorption in haemoglobin in the green and orange wavebands explain why the effect of krypton green, KTP, copper vapour and certain continuous wave dye lasers have similar effects on vascular lesions. (KTP stands for potassium tylenol phosphate, the material of a crystal inserted into a Nd:YAG laser to halve its wavelength to 532 nm).

The appropriate laser must be chosen for each application.

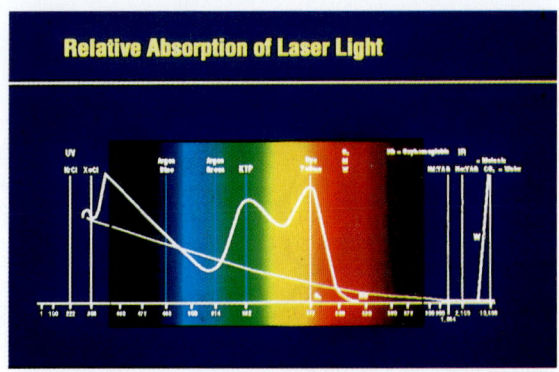

Fig. 4.10

Ophthalmology

Various lasers are in use for all types of eye surgery. The eye is largely transparent to visible and near infra-red radiations which are not absorbed by water, while the vascularised and pigmented retina does absorb these wavelengths. Diabetic patients often develop increased vascularisation of the retina leading to loss of vision when these vessels bleed or invade the fovea. Blood vessels can be sealed using the argon or krypton laser depending upon their position on the retina. Retinal tears are usually treated by surrounding them with a ring of small laser burns.

Radiation from the ruby or Nd:YAG laser penetrates the cornea but is partially absorbed in the pigmented iris, so can be used for making the incisions used to treat glaucoma. The Nd:YAG laser is also used to puncture the lens capsule when it regrows after cataract surgery.

Short wave ultra-violet radiation does not penetrate into the eye, so that excimer lasers with UV emission lines can be used for sculpting the cornea as a cure for myopia.

Figure 4.11, demonstrates the penetration depths.

Excimer Lasers

These derive their name from "excited dimer", where a dimer is a molecule containing two atoms. In the excited state the molecule is

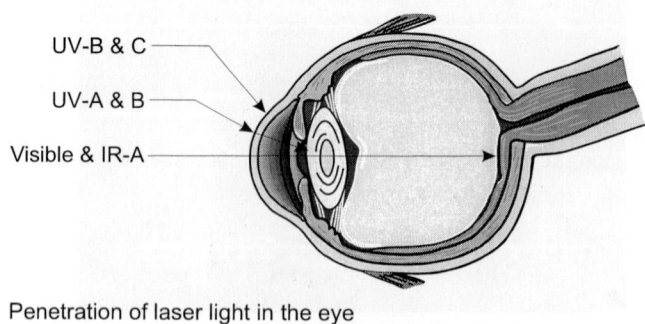

UV-B & C

UV-A & B

Visible & IR-A

Penetration of laser light in the eye

Fig. 4.11

bound together but, as it decays to the ground state with the emission of a photon, the atoms repel each other so that the molecule splits. This process leads to rapid population inversion. Inert gas excimers using xenon and argon with fluorine, emit in the UV-B and UV-C regions.

ENT Surgery

The CO_2 laser is used for the destruction of nodules and polyps on the vocal cords and for the removal of recurrent respiratory papillomas. In the latter condition the larynx is filled with viral warts which disturb breathing and alter the voice. Laser surgery ensures removal without bleeding or oedema. Lesions can be removed from the tongue, tonsils and buccal mucosa, and snoring can be prevented by laser surgery to the uvula.

Surgery in the middle ear requires fibre optic transmission because of the very limited access. Argon and Nd:YAG lasers are therefore used.

Gastroenterology

The Nd:YAG laser is used via an optical fibre in an endoscope to burn away cancerous tissue in the oesophagus, it relieves obstruction although it does not normally eradicate the malignancy. Pre-malignant cell changes at the gastro-oesophageal junction (Barrett's Oesophagus) can be cured with photo-dynamic therapy. Argon or Nd:YAG lasers are effective to coagulate the tissue around bleeding peptic ulcers, to remove small malignant stomach lesions and to reduce the bulk of larger ones.

Stones in the gall bladder (and kidney) can be shattered by the photo-acoustic action of a laser. The small fragments may then be passed naturally or removed endoscopically. However, not all stones respond as their composition is not uniform.

Photodynamic therapy can be successful in the treatment of bile duct obstructions.

Gynaecology

CO_2 lasers are valuable for the removal of vaginal polyps and pre-cancerous lesions of the cervix. Menorrhagia and endometriosis can be treated with Nd:YAG irradiation. Other applications include the treatment of adhesions and polycystic ovaries.

Dermatology

Port wine stains, which are present because of excess blood vessels below the epidermis, respond to argon laser treatment. A Q-switched dye laser at 577 nm may also be used.

A dye laser emitting at 630 nm in conjunction with an appropriate photosensitising drug is employed in PDT for the pre-malignant skin lesions of Bowen's disease. Diode lasers may be used but, because of the low output, treatment times may be unacceptably long. Diode laser output is increasing with new construction techniques. Broad band red light sources, 580–720 nm, are also used for PDT of Bowen's lesions.

CO_2 lasers can been used for the removal of tattoos, but radiation from the Q-switched ruby laser at 694 nm is minimally absorbed in haemoglobin, so gives less scarring. They are used cosmetically to remove facial wrinkles.

Dye Lasers

The dye laser was developed to allow for variation of wavelength within one lasing cavity. The liquid in the cavity contains an organic dye which can be changed by flushing the system and refilling it.

The yellow wavelength 577 nm is used for coagulation and vein occlusion, but different skin pigmentation can be treated more effectively with other wavelengths.

In addition to the dermatological applications suggested above, red light from a dye laser can treat arterial blockages more satisfactorily than by using a balloon catheter. Unfortunately, treatment with a balloon which is inflated at the site of a blockage is unsuccessful in about one third of cases which rapidly block again. Photodynamic therapy seems to suppress regrowth of arterial muscle cells.

Physiotherapy

Low power He:Ne lasers and semi-conductor diodes have a place in physiotherapy to supply heat to discrete small areas of tissue. Conditions as diverse as muscle injuries, joint swelling and pressure sores respond to this treatment. Power output can be increased with a cluster probe of several diodes mounted in a circular array.

Lasers for Diagnosis

Visible and infra-red radiation for diagnosis is covered in Chap. 6.

For more information about the use of medical lasers the reader is referred to Refs. (10), (17) and (18).

Laser Safety

Lasers are grouped into classes according to the degree of hazard. Classes 1, 2, and 3A are safe for viewing with the unaided eye because their maximum power and irradiance are limited. If the emission is in the visible range protection is afforded by aversion responses such as the blink reflex. Class 3B and 4 lasers present serious hazards and must not be used without taking appropriate precautions.

All the lasers mentioned in this section, with the exception of some He:Ne devices are class 4 lasers. Eye protection must be worn, local rules are required for safe use and personnel must receive training before being allowed access. Examples of class 2 lasers encountered in hospitals are laser pointers and isocentre defining lights on radiotherapy equipment. Occasionally these devices may be class 3A.

Full details of laser hazards and safety precautions are included in the UK document: "Guidance Notes on the Safe Use of Lasers" (19). Similar documentation has been issued in the EC and in the USA, and several books are also available, (20). Guidance Notes should always be consulted before using medical lasers. Where hazardous lasers are used, it is normal for a laser protection adviser to be appointed.

Microwave and Radiofrequency Radiation

At wavelengths greater than the infra-red band, i.e. more than 10^6 nm in the electromagnetic spectrum, it is usual to refer to frequency rather than wavelength. Microwaves, which are electromagnetic waves from 1 mm to 1 m in wavelength, are equivalent to the frequency range 300 MHz to 300 GHz, and radiofrequency radiation is from 100 kHz to 300 MHz. These radiations are usually produced from a magnetron, which has been described in Chap. 3 (pages 75 & 76), and are used in the medical field to produce heat at depth in tissue. Figure 4.12, shows the depth of penetration of various frequencies.

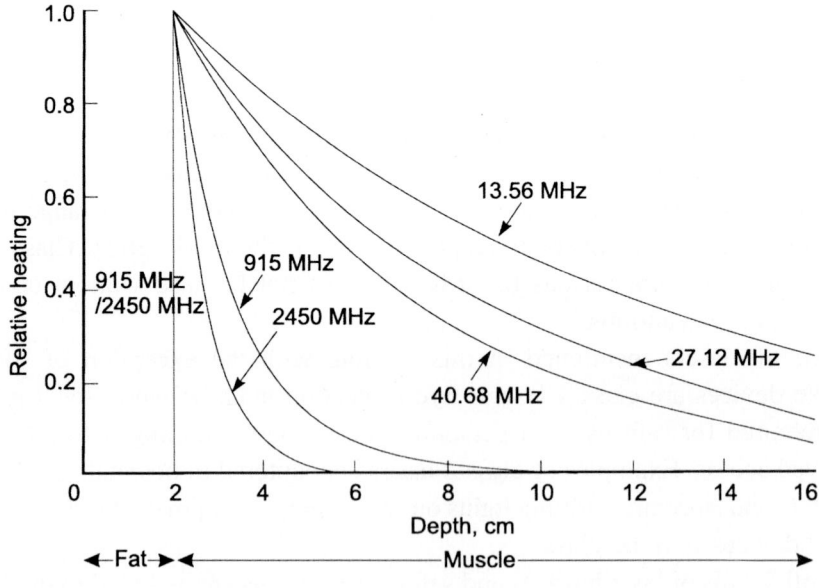

Fig. 4.12

When magnetrons were first used in radar applications in World War II, it was noticed that objects placed near to them became hot. Microwave heating was first patented for cooking in the USA in 1945 but microwave ovens did not arrive in the UK in numbers until the early 1980s. Microwave ovens are now the most common microwave sources universally. The principle of operation is simple. A magnetron generates microwaves at an operating frequency of 2450 MHz and these are conducted into a metal walled cooking chamber by a waveguide.

Tissue Heating

Aches and pains have been treated with heat since ancient times. Heat applied to the skin causes an increase in blood flow, by as much as five times the normal rate at a temperature of 42°C. There is also an increase in hydrostatic pressure in the capillaries and an increase in active capillary surface area. Heating increases the heart rate, allowing efficient flushing away of waste products as more capillary beds open.

At normal body temperature (37°C), collagenous tissue in ligaments, tendons and capsules has high tensile strength and minimal viscous flow. Heating increases viscous flow and relaxes tension resulting in some elongation of tissue. These effects occur up to about 44°C. At about one degree higher than this there is a danger of tissue destruction.

Heat is effective in the relief of muscle pain, but the mechanism is less well understood. It is believed that, when the brain perceives two different sensory stimuli, one takes precedence over the other. When heat and pain are received simultaneously the perception of pain is reduced in proportion to the increase in temperature.

The first person to experiment with tissue heating was the French physiologist, A. d'Arsonval in Paris in 1890 (21). He passed an electric current through a frog's muscle and noticed that, as the frequency of the current increased, the contraction of the muscle reduced until, at about 10 Hz, movement ceased altogether. On applying the current to himself, he noted a pleasant warm sensation and reddening of the skin with increased blood flow. Several French doctors tried applying electric currents to their patients and said that it was an effective treatment for a vast range of conditions.

Fig. 4.13. Local autoinduction cages.

A cumbersomely-named treatment, d'Arsonvalisation, was introduced. A limb or larger portion of the patient was placed in the electric field of a solenoid, (Fig. 4.13). An added gimmick was for the operator to produce sparks from a part of the patient under treatment.

In 1895 in England, the Society of Trained Masseuses was founded and began to provide training for its members. An announcement in their 'Nursing Notes' advertised: "Massage and Electricity scientifically taught in a fortnight. Lessons on living subjects daily."

In 1908, Franz Nagelschmidt in Germany, demonstrated the deep heating effect of high frequency current and named the process 'diathermy' which was derived from the Greek for 'through heat'. This was to distinguish it from radiant heat which is mostly absorbed in the first 1 mm of surface thickness.

Diathermy provided Induced Fever Therapy which was used to treat various conditions ranging from venereal disease to pneumonia. It became especially fashionable following the successful treatment of King George V for pneumonia in 1929.

Early long wave diathermy operated at about 1 MHz but it was difficult to contain the radiation. In 1928 Jena produced a 3 m 400 watt equipment for the first short wave diathermy. Short wave diathermy today normally operates at 27 MHz (11 m wavelength).

Short Wave Diathermy

Physiotherapists first began treatments using short wave diathermy in the late 1920s. Tissue heating is almost entirely due to currents produced in conductive materials and is similar to the heating effect in the element of an electric fire where the resistance to the current is due to the impedance offered to the flow of electrons by the molecular structure of the material. In the case of diathermy the electron flow is set up by the high frequency field and, as the electrons collide with the molecules in the tissue, some of the energy is converted into heat.

The part of the body or limb to be treated is coupled to the circuit either capacitatively or inductively. In the capacitor field technique, two plates connected to the diathermy output are placed on either side of the limb. The plates have convex edges to produce a uniform electric field pattern and their position can be adjusted within plastic covers. The induction coil technique employs a cable coiled round the limb or the cable is coiled into a helix which is placed close to the part of the body to be treated. Alternating current through the coil produces a magnetic field in nearby tissue. This sets up eddy currents in the tissue which produce the heating effect.

Surgical Diathermy

High intensity diathermy has the property of sealing blood vessels and is employed for cutting tissue. A current is applied through an electrode where an arc is set up and causes localised heating of the tissue. A large-area return electrode is placed under the patient in contact with the skin surface.

Microwave Therapy

High frequency waves from a microwave generator pass into the body and interact with tissues with high water content. Water consists of polar molecules with a single positive and a single negative charge. The microwaves cause the molecules to vibrate and some of the energy is converted into heat.

Microwave appliances usually operate at 2450 MHz. The oscillating R-F field is fed through a coaxial cable to an aerial, usually called an applicator, positioned some 12 to 15 cm from the body. Power levels are adjusted so that the patient feels comfortably warm, usually at between 20 and 80 W.

Microwave Hyperthermia

Malignant tumours usually have poor oxygenation because the blood supply to a tumour is less good than that to surrounding normal tissue. This means that heat is not dispersed efficiently. Some tumours can be successfully treated by targeted microwave heating, sometimes in conjunction with ionising radiation. The aim of hyperthermia is to raise the temperature of the tumour to about 43°C to cause cell destruction, and to increase the blood supply through vasodilation to make remaining malignant cells more radiosensitive.

Whole body hyperthermia has been used for the treatment of widespread cancerous deposits in association with radiotherapy. Although producing encouraging results, it does need very careful monitoring because the heat induced increases in pulse rate and cardiac output.

Hazards

Microwave and R-F radiations generate heat which is removed by the blood supply, so the main areas at risk are organs with low vascularity. Most at risk is the eye because the lens contains a protein called crystalin which is similar to albumin found in egg yolks. It turns white when heated and cataracts are formed. Power densities greater than 100 mW cm^{-2} can produce cataracts.

Various other unpleasant biological effects have been reported as resulting from exposure to these frequencies, but none has been proved.

International standards for human exposure have been formulated (22), and all microwave and R-F radiating equipment should be checked and have its operation assessed by a competent person.

References

1. Lerman S., *Radiant Energy and the Eye*, Bailliere Tindall, London, 1980, ISBN 0-02-369970-1.
2. Smith K. C., *The Science of Photobiology*, Plenum Press, New York, 1977.
3. Freund L., *Phototherapy*, in *Elements of General Radiotherapy*, Rebman, London, 1904.
4. Wright C. A., *Notes on Instrumentation for Radiotherapy*, Rebman, London, 1904.
5. Alderson H. E., *Arch. Dermatol. Syphilol.* **8**, 79–80, 1923.
6. El Mofty A. M., *Vitiligo and Psoralens*, Pergamon Press, New York, 1968.
7. Van Weelden H. and De La Faille H. B., *British J. Dermatol.* **19**, 11–19, 1988.
8. Parrish J. A. *et al.*, *UV-A, Biological Effects*, John Wiley, Chichester, 1978, ISBN 0-471-99759-5.
9. Diffey B. L., *Ultraviolet Radiation in Medicine*, Adam Hilger, Bristol, 1982, ISBN 0-85274-535-4.
10. Moseley H., *Non-Ionising Radiation*, Adam Hilger, Bristol, 1988, ISBN 0-85274-166-9.
11. Diffey B. L. (ed.), *Radiation Measurement in Photobiology*, Academic Press, London, 1989, ISBN 0-12-215840-7.
12. Gordon J. P., Zeiger H. J. and Townes C. H., *Phys. Rev.* **99**, 1264–1274, 1955.
13. Schawlow A. L. and Townes C. H. *Phys. Rev.* **112**, 1940–1949, 1958.
14. Maiman T. H., *Nature*, **187,** 493–494, 1960.
15. Bridges W. B., *Appl. Phys. Lett.* **4**, 128–130, 1964.
16. Patel C. K. N., *Appl. Phys. Lett.* **7**, 15–17, 1965.
17. Carruth J. A. S. and McKenzie A. L., *Medical Lasers*, Adam Hilger, Bristol, 1986, ISBN 0-85274-560-5.
18. Fuller T. A., *Lasers in Surgery and Medicine* **1**, 5–14, 1980.
19. *Guidance on the Safe Use of Lasers in Medical and Dental Practice*, Medical Devices Agency, London, 1995, ISBN 1-85839-488-0.

20. Sliney D. H. and Trokel S., *Medical Lasers and their Safe Use*, Springer Verlag, Heidelberg, 1992, ISBN 3-540-97856-9.
21. D'Arsonval A., *Proc. Société de Biologie,* Feb. 1891.
22. IRPA, *Guidelines on Limits of Exposure to Radiofrequency Electromagnetic Fields in the Frequency Range 100 kHz to 300 GHz,* Health Physics, 1984.

Chapter 5

NUCLEAR MEDICINE

Nuclear medicine covers the medical uses of radioactive isotopes primarily for diagnostic purposes but also, in a few instances, for therapy. For diagnosis, the aim is to be able to make a clinical diagnosis while giving the least practicable radiation dose to the patient. Diagnostic procedures can be divided into four general categories which depend upon either, localisation, dilution, diffusion (or flow), or biochemical and metabolic properties. Diagnostic radiology and nuclear medicine are complementary techniques. In general, diagnostic X-rays show body structure whereas isotope studies show function.

Diagnostic nuclear medicine relies upon the use of artificially produced isotopes with short half lives in order that patient dose is minimised, and that repeat investigations are possible. The production of such isotopes, although not the very short lived ones used today, began in 1932 when the British physicists J. D. Cockroft and E. T. S. Walton built a high voltage particle accelerator capable of producing protons with sufficient energy to cause nuclear transformations.

In the first Cockroft–Walton experiments, lithium nuclei, containing three protons, were bombarded with protons. Those nuclei which captured a fourth proton from the beam were transformed into beryllium nuclei which themselves split into two helium nuclei (alpha particles).

Rutherford's classic experiments at the Cavendish Laboratory, Cambridge, in 1919, had established that alpha particles were capable of transforming one element into another. He used radium alpha particles to convert stable nitrogen into radioactive oxygen-15 which has a half life of two minutes.

Shortly after the Cockroft-Walton accelerator was built, E. O. Lawrence at the University of California in America, developed a circular accelerator called the cyclotron. This used a magnetic field combined with a rapidly oscillating voltage to accelerate nuclear particles along a spiral path, thus taking up less space than a voltage multiplier. Very high energies were eventually achieved with the cyclotron.

Early Experimental Work

The first work in diagnostic nuclear medicine was described by George Hevesy, a Hungarian born physicist working in Denmark. In the early 1920s, he grew bean plants in water containing a known amount of naturally occurring radioactive lead. By removing leaves regularly and analysing their radioactive content, he was able to trace the course of the lead through the plant. His first animal experiments in 1935, (working with D. Chiewitz), involved feeding phosphorus-32 labelled sodium phosphate to laboratory rats and measuring the distribution in body parts at autopsy after 20 days. (1).

Clinical Diagnosis with Radioactive Isotopes

In 1949, reactor-produced iodine-131 with a half life of 8.04 days, was first used by N. Veall, in London, to diagnose abnormalities of the thyroid gland. (2). The rate at which ingested or injected I-131 accumulates in the thyroid, to be metabolised into the hormone thyroxine, and the fraction of the administered dose collected, are measures of the gland's function.

Unfortunately, the principal gamma emission from I-131 is at 364 keV, and early detection methods relied upon the Geiger-Muller counter. At this high energy, most of the gamma rays pass through a G-M tube without being absorbed, resulting in a very low detection rate. Veall constructed a collimated G-M tube and succeeded in mapping the spatial distribution of iodine in the thyroid.

The dose of iodine-131 was up to 2 mCi (about 70 MBq in S.I. units), which exposed the patient to a dose which would be unacceptably high by modern standards.

Another major experimenter in early nuclear medicine was W. V. Mayneord, physicist at the Royal Cancer Hospital (now the Royal Marsden) in London. In 1950, he edited a booklet on this new modality (3) which described, in addition to thyroid investigations, the use of phosphorus-32 to estimate red cell volume, and of sodium-24 to study muscle clearance. There was also a description of the use of carbon-14 labelled compounds for carbon dating.

The Radiochemical Centre at Amersham, England, (now renamed Nycomed Amersham) was established to supply radioactive isotopes. Tracer techniques began to be used more widely.

Scintillation Counters

An important breakthrough was the introduction of scintillation detectors. When ionising radiation passes through a scintillator, a light pulse is emitted whose intensity is proportional to the energy given up by the radiation to the scintillator. Sodium iodide, activated by the addition of a small quantity of thallium, is one of the best scintillators for the detection of gamma radiation, and this material is used in almost all nuclear medicine equipment, although early work in America by B. Cassen and in England by R. J. T. Herbert used calcium tungstate as the scintillator (4), (5).

In sodium iodide, the photoelectric effect is dominant at energies below 500 keV, Compton scattering between 500 keV and 4 MeV, and pair production at higher energies.

When a photomultiplier tube is coupled optically to the NaI crystal, the light output of the crystal is converted to an electrical pulse whose amplitude is proportional to that of the light pulse, and hence to the detected amount of radiation.

A major difficulty in imaging gamma rays is that they cannot be focused like X-rays, and other methods of determining the position of origin of each ray must be found. In 1949, Copeland and Benjamin had used a pinhole collimator to produce images of the gamma rays from radium needles, but, inevitably, the sensitivity of such detectors is low making a pinhole unsuitable for diagnostic work.

Mayneord and Cassen developed the first lead collimating devices for use with scintillation crystals in order to produce rectilinear scanners. The collimated scintillation counter moved, automatically, backwards and forwards across the region to be scanned to build up a map of the amounts of activity in each small section (Fig. 5.1). This was a slow process, requiring the patient to remain still throughout, but nevertheless, this technique was in common use until the 1970s.

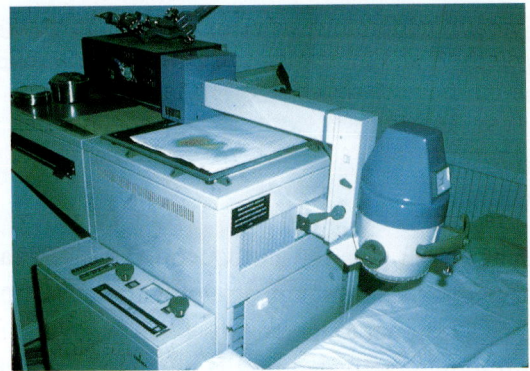

Fig. 5.1

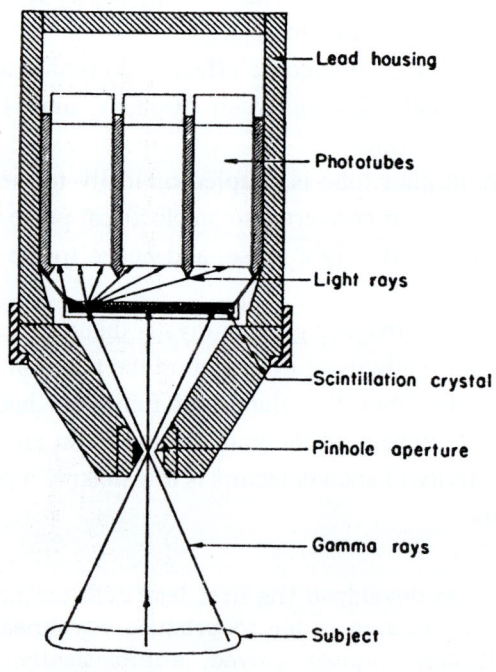

Sectional drawing of scintillation camera (1957).

Fig. 5.2

The Gamma Camera

The 'spot by spot' technique of the rectilinear scanner was wasteful of time, and various workers felt that it should be possible to design a stationary detector which could produce an image of the radioactivity in a whole region of the body simultaneously. This was first achieved in the Donner Laboratories in California in 1957 by Dr H. O. Anger and colleagues (6). Anger's first gamma camera used a four inch diameter crystal of thallium-activated sodium iodide which was viewed by an array of seven photomultiplier tubes. Each photon detected by the crystal produced a pulse in each of the photomultiplier tubes, and the relative pulse sizes were used to compute the position of each photon. Figure 5.2, shows an illustration which accompanied Anger's first article.

Parallel-hole collimators which reduced geometric distortion and improved sensitivity, and larger crystals which increased the field of view and incorporated 37 photomultipliers arranged in three concentric hexagonal rings, were provided before the first commercial Anger camera became available in the early 1960s. A typical collimator was 25 mm thick and perforated by up to 25,000 parallel holes, allowing only those gamma rays travelling at right angles to the crystal surface to be detected. These refinements produced sharp images with minimal blurring.

Crystalline sodium iodide is hygroscopic, so the crystal was supplied in a thin walled aluminium case with a glass window at the end facing the photomultipliers. The aluminium had the added advantage of cutting out low energy beta radiation which would otherwise cause fog in the image.

Competitive techniques emerged in the late 1960s and early 1970s. Multicrystal gamma cameras used a detector matrix of 294 separate crystals connected to a computer (7). They offered the advantage of higher count rates, but had poor resolution and they were very expensive. They do still have application in PET scanning (see p. 141).

A solid state camera using a high-purity germanium detector was constructed but, although producing good pictures, it was prohibitively expensive.

Instruments relying upon gaseous detection, such as the spark chamber, were tried but had poor sensitivity.

Artificial Radioisotopes

The next advance in nuclear medicine occurred in the mid 1960s with the introduction of monoenergetic gamma emitters with short half lives which could be prepared close to the nuclear medicine department where they would be used.

Many isotopes are produced by the nuclear decay of a radioactive isotope, and a daughter isotope may be stable or radioactive. For some parent isotopes, the decay process is in two stages. Firstly, there is the emission of a beta particle (electron) producing an excited daughter nucleus, and then, after some delay, there is the emission of a gamma ray.

The excited state of the daughter nucleus is referred to as a metastable state, and is denoted by writing the letter 'm' next to the mass number. The change from metastable to stable state is by the emission of a pure gamma ray after a relatively short time (which gives a half life to the metastable radionuclide). These two conditions make such isotopes ideal nuclear medicine agents. Generators containing the parent isotope can be installed in radiopharmacies and 'milked' at intervals to provide the metastable daughter. It is an added benefit if the parent isotope has a relatively long half life.

Two isotopes were favoured initially. Molybdenum-99, with a half life of 67 hours, decays to metastable technetium-99m, which, with a half life of just over 6 hours, itself decays to stable technetium-99. The monoenergetic gamma ray involved in this decay is at 140 keV.

Tin-113 has a half life of 118 days decaying to metastable indium-113m whose half life is 1.7 hours. Decay by the emission of a gamma ray at 390 keV produces stable indium-113.

Indium-113m has the advantages of a longer lived parent, making it possible to use one generator for up to three months, and a short half life itself. The greater penetrating power of the gamma rays was also considered to be an advantage in the early stages, in that deep seated processes might be more clearly imaged. However, heavy collimators are needed to eliminate scattered radiation, and, unfortunately, it was much more difficult to produce labelled pharmaceuticals with indium-113m than with technetium-99m. For routine work, technetium-99m became the isotope of choice for almost all investigations.

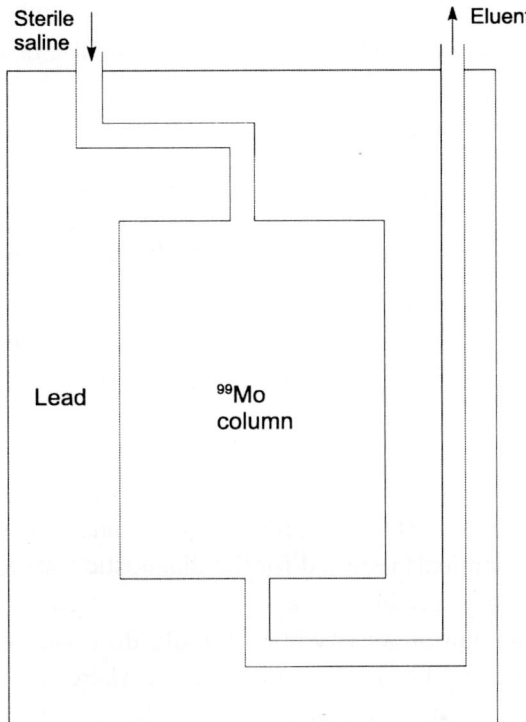

Sterile saline

Eluent

Lead

^{99}Mo column

Fig. 5.3. Schematic diagram of ^{99m}Tc generator.

Figure 5.3, shows a schematic diagram of a technetium-99m generator. The technetium is eluted (washed off) as sodium pertechnetate by forcing sterile saline through the generator using a vacuum vial. The molybdenum is more firmly attached to the alumina column and remains behind so that the generator can be eluted repeatedly for about a week before the decay of the parent isotope leads to too little technetium being available. The generator is shielded by lead casing and the eluate vial must also be protected.

Technetium-99m can be used in a wide selection of nuclear medicine studies because it is relatively easy to label many different compounds and colloids. Labelling usually involves simply mixing or shaking the pertechnetate with the compound to be labelled, and any other necessary chemicals, at room temperature, although some labelling involves incubation in a water bath. Being chemically similar to iodine, Tc-99m can also be used for thyroid studies.

Scanning with Technetium-99m

The stages in the process of diagnosis using technetium-99m and the gamma camera are:

1. Preparation of the radionuclide by elution of the generator. Typically, this takes place at the beginning of each working day. In accordance with the Ionising Radiations Regulations, the generator is housed in a lead-protected cabinet which must also obey the Medicines Act Regulations on sterility. Quality Control checks are routinely carried out to ensure that the technetium is not contaminated with molybdenum.
2. The pharmaceuticals required for the diagnostic tests to be undertaken must be labelled. Labelling takes place under sterile conditions behind lead. The amount of activity in each multi-dose vial of radiopharmaceutical is measured in a 'well' type counter where an ionisation current is produced by the gamma rays from the sample. This current is proportional to activity but also depends on the gamma energy. Most well counters for radionuclides are automated so that it is only necessary to enter the radionuclide on the control panel, and activity in MBq then appears on the display. The dose calibrator must be checked on a regular basis with a long lived sample source such as caesium-137, (half life 30 years), and the labelled compounds are checked for sterility, chemical purity and the absence of free pertechnetate.
3. The radiopharmaceutical is drawn up into a shielded syringe and the activity in the patient's dose is checked, before injection into the patient. After an appropriate interval the area or organ to be investigated is positioned in front of the gamma camera.
4. Scintillations in the sodium iodide crystal are converted to electronic signals and the image is displayed. Computer smoothing may be applied and the processed image then displayed for diagnosis to be made.

Some Planar Scanning Techniques

Bones

Phosphate compounds are bone-seeking agents. The most commonly used agent is technetium-99m labelled diphosphonate which does not bind to red blood cells as most phosphates do. The average activity used for a skeletal study is about 550 MBq giving a whole body dose of about 5 mSv to the patient. This study is chiefly useful in the diagnosis of metastatic malignant disease and fractures which do not appear clearly on radiographs such as stress fractures. Figure 5.4, shows two whole body bone scans. The one on the left is normal while that on the right shows bony metastases.

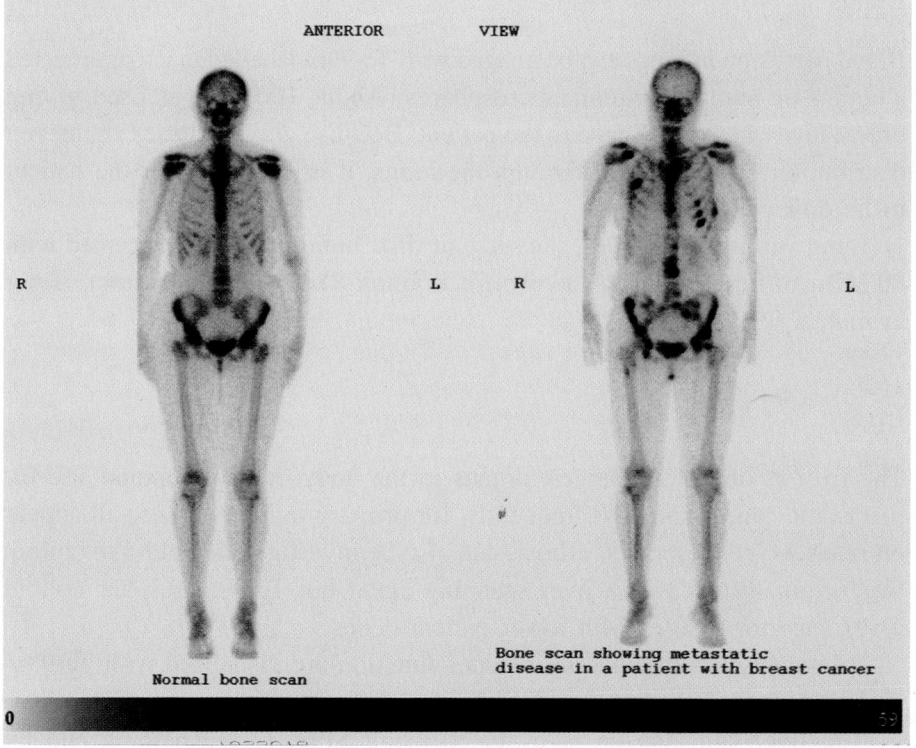

Normal bone scan

Bone scan showing metastatic disease in a patient with breast cancer

Fig. 5.4

Brain

The blood-brain barrier prevents most materials from being transferred from the blood to neural tissues. Some diseases cause a breakdown in this barrier, and this is particularly true of brain tumours. Approximately 550 MBq of technetium-99m pertechnetate can be used for planar scanning some three to four hours after injection. More rapid scanning is possible if Tc-99m diethylenetriamine pentaacetic acid (DTPA) is used.

Planar scanning has recently been replaced in many centres by SPECT scanning and magnetic resonance imaging the latter gives particularly good images of brain tumours.

Lungs

Blood perfusion in lungs can be imaged with Tc-99m labelled macroaggregated albumin or human albumin microspheres. About 100 MBq is used giving only a low whole body dose to the patient. Because gravity causes an uneven distribution of blood flow through the lungs, it is important for the patient to lie down if possible.

Lung ventilation studies, showing air distribution, can be performed with 80 MBq of Tc-99m DTPA aerosol, if krypton-81m gas, (see below), is not available.

Liver

The liver is one of the largest organs in the body. It is a common site for metastatic cancer and, less frequently, for primary tumours. These all appear on scans as 'cold' spots. Cirrhosis can also be investigated. Gold-198 colloid was originally used as a liver scanning agent but Tc-99m sulphur colloid gives superior images with lower patient doses.

Bilary obstruction and gall bladder function are visualised with Tc-99m dimethyl iminodiacetic acid, (HIDA).

The liver can also be investigated using SPECT (see below), and by ultrasound, (see Chap. 6), which is usually the recommended method when it is desirable to avoid radiation exposure.

Other planar techniques include kidney imaging to investigate structure and relative function. The reader is referred to nuclear medicine text books, e.g. (8), (9).

Gamma Camera Quality Control

Regular checks must to be made on field uniformity. This is done using a 'flood-field' phantom which is positioned in front of the collimator. Some are designed to be refillable with Tc-99m as required, while others are solid Co-57 sources. Co-57 has a similar gamma ray energy of 0.122 MeV, but a long half life of 270 days.

Spatial resolution and linearity are checked using bar phantoms containing attenuating materials in various geometric configurations, while contrast resolution is investigated using radioactive solution in perspex containers. Energy resolution is controlled by a pulse height analyser, the settings of which must also be checked regularly.

For further information on nuclear medicine quality control, a good starting point is Ref. (10) which, although not modern, provides a full account of the basic principles.

Dynamic (Time Varying) Imaging

Information can be obtained on rates of uptake and sites of accumulation of radio-pharmaceuticals by recording a series of images from the patient. This usually starts immediately after administration of the dose, and is an extremely useful test of organ function.

One of the earliest dynamic studies was renography, in which time-activity data from both kidneys, the ureters and bladder can be recorded to determine whether the excretory system is functioning normally or abnormally.

The earliest work used diuretics which, at the time, were usually compounds of mercury and the drug itself incorporated radioactive mercury. Mercury-203, with a half life of 47 days, was tried first, followed by mercury-197 whose half life was 65 hours (11).

Over the years techniques have been refined. Tc-99m labelled Hippuran is a 'urine seeking' agent which became popular. Nowadays, various different compounds are used for the different types of study. Tc-99m DTPA is effective for clearance studies and gives a low dose to the patient. Tc-99m MAG-3 gives even lower doses and is always used for children.

Modern gamma cameras allow the operator to outline areas of interest on the computer screen using a cursor. The total number of counts within the outlined area is measured on each of a series of images and the computer produces graphs of rates of uptake and excretion (Fig. 5.5).

The functioning of the beating heart can be similarly studied. A rapid series of images is required in this case and a cine-loop is prepared which can be displayed repeatedly on the screen. Triggered by the R-wave from an

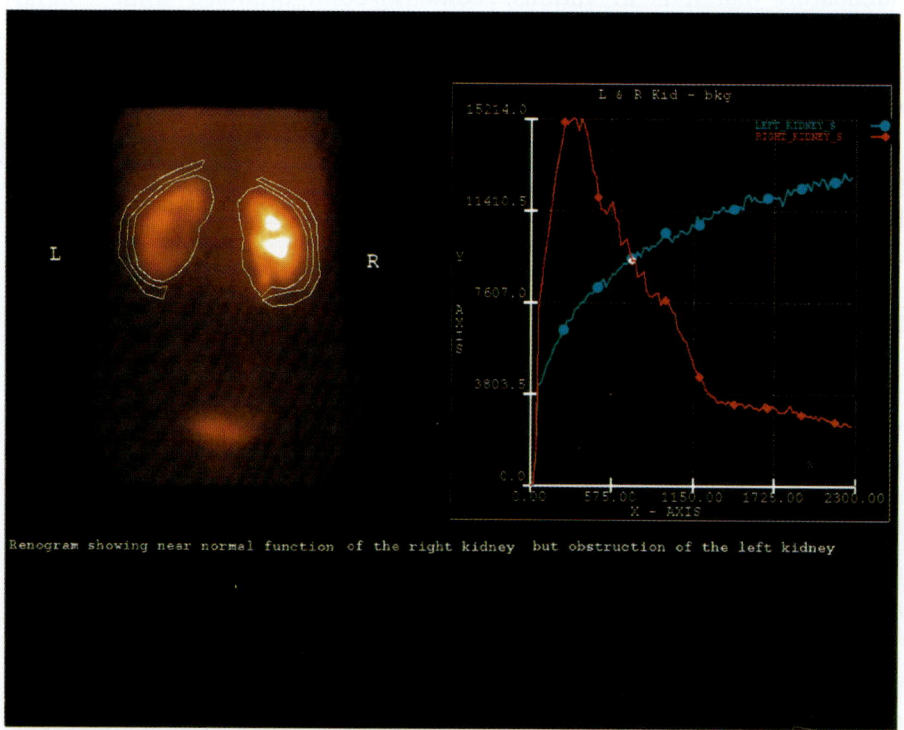

Renogram showing near normal function of the right kidney but obstruction of the left kidney

Fig. 5.5

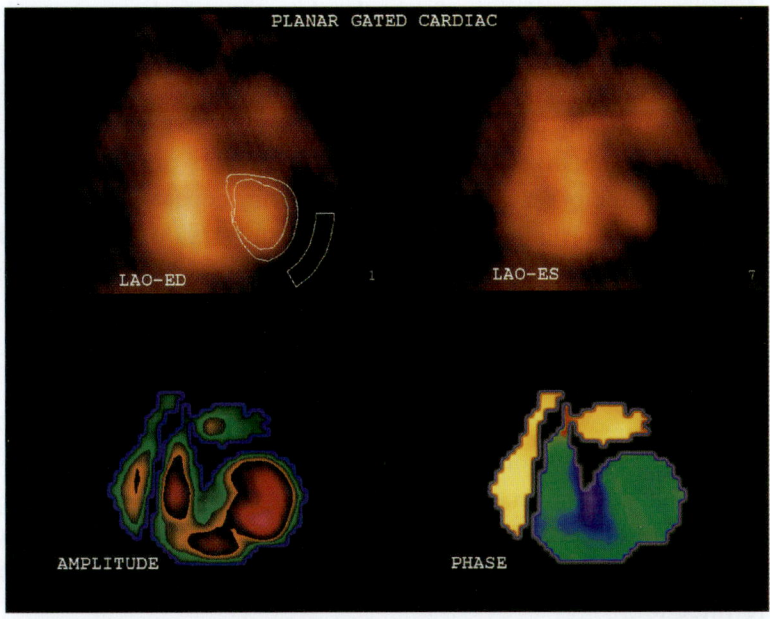

PLANAR GATED CARDIAC

LAO-ED   1   LAO-ES   7

AMPLITUDE   PHASE

Fig. 5.6

electrocardiogram, some 20–30 frames, each lasting about 40 ms, are acquired and augmented by the gamma camera's computer. Repeated display of the sequence showing the pulsating heart is analysed to show the efficiency of the chambers of the heart at pumping blood, the ejection fraction, and other quantitative data. The relative sizes of the chambers are measured and defects in coronary blood flow can be demonstrated (Fig. 5.6).

Cardiovascular disease is the leading cause of death in the western world and much use is made of cardiac scanning in its diagnosis. See also the section of tomography.

Other Radionuclides and Their Application

- Gallium-67, a cyclotron produced isotope, emits gamma rays of three main energies, its half life is 78 hours. Gallium-67 citrate binds to plasma proteins and is useful in the detection of abscesses and other inflammatory lesions. It can be used in the detection of lymphatic cancers including Hodgkin's disease and has been used for staging lung cancer. Many patients with Acquired Immunodeficiency Syndrome (AIDS) can be investigated, where 'hot' spots are seen at sites of pulmonary and other infective areas. Primary hepatoma, an uncommon liver cancer, normally shows up as a 'cold' area on Tc-99m sulphur colloid scans, but it does take up gallium-67 allowing the confirmation of diagnosis.

- Indium-111 emits gamma rays of 173 and 247 keV. It is produced in a cyclotron and has a half life of 67 hours. It is used to label white cells for the detection of sites of infection and abscesses (as in Krohn's disease). It is also useful for the labelling of platelets in the location of thromboses and to test for the rejection of transplanted organs. The labelling of blood constituents is a time-consuming and intricate process since cells must not be damaged either during extraction, labelling or re-injection.

- Krypton-81m with a half life of 13 seconds, requires a generator containing the parent isotope rubidium-81 whose half life is only 4.7 hours. The generator is useful for less than a day so presents transport problems anywhere other than in easy reach of a cyclotron. Kr-81m is an inert gas emitting 190 keV gammas. The generator is eluted with compressed air and the patient inhales the eluate for lung ventilation studies. When used in conjunction with a technetium labelled HSA macroaggregate for lung perfusion imaging, it is possible to detect pulmonary embolism with a gamma camera capable of discriminating between the two gamma energies.

- Thallium-201 is cyclotron produced with a half life of 73 hours. It decays by electron capture emitting 80 keV X-rays. It is chemically similar to potassium, and can be used as thallous chloride for myocardial perfusion imaging and for various studies of the central nervous system.

- Xenon-133 is produced in a nuclear reactor. It emits beta particles and low energy (81 keV) gamma rays. It is an inert gas and is used, with re-breathing, in lung ventilation imaging.

Tomography

The planar imaging so far described produces two dimensional information. The acquisition of pairs of images at right angles allows for three-dimensional location of sites of activity using geometrical reconstruction, but there are problems with this method because of the presence of activity in areas surrounding the region of interest. The development of X-ray computed tomography gave the possibility of a similar reconstruction in nuclear medicine.

Nuclear medicine tomography was first undertaken in Philadelphia in 1964 by D. E. Kuhl, (12), who used an analogue technique. The first digital CT system for nuclear medicine was built in Aberdeen, Scotland in 1968 by A. R. Bowley and colleagues (13). This had its first application in the confirmation of the diagnosis of epilepsy.

To obtain tomographic data it is necessary to produce images of radioactivity distribution from a large number of orientations. The devices involved can be divided into those systems which operate in the same manner as early X-ray tomography and blur out unwanted planes giving a generally near constant background, and those which reconstruct the distribution more fully by methods similar to x-ray CT scanning (SPECT).

The former, called focal plane tomography, was first applied with the rectilinear scanner using a wide-angle collimator with short focal length. For any particular position of the detector, radioactivity in only one small area of the focal plane would be detected, but for other planes points of activity would appear as circles of relatively greater diameter leading to blurring.

The technique can be extended to the gamma camera by the use of a focused collimator, as in Fig 5.7, or by attaching a rotating slant-hole collimator. With the collimator holes inclined at some 25° to the axis of rotation, circular images of point sources will appear on the display screen. The radii of circles is proportional to the distance from the face of the collimator. The computer can then reconstruct a longitudinal tomographic image.

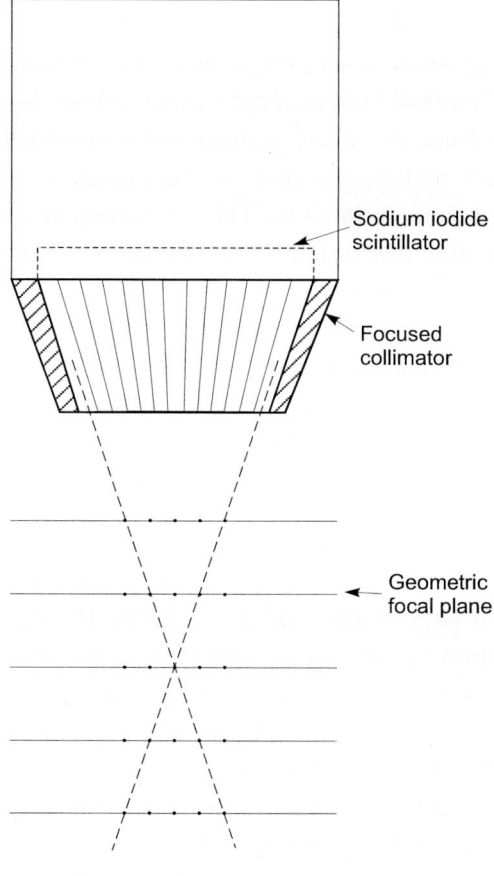

Fig. 5.7

SPECT

The technique now known as SPECT (single photon emission computed tomography) became widely available in the early 1980s following the increased availability of digital systems. SPECT acquires data from all around the object and provides superior results. The gamma camera must be mounted on a ring which will allow complete rotation.

In the 'step-and-shoot' method, the camera stops to acquire one planar image then rotates through a certain angle, typically 6°, before acquiring

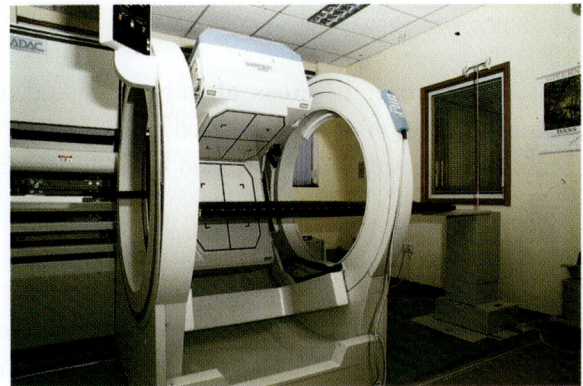

Fig. 5.8

another image. The process is repeated around the circle. The disadvantage of this method is that time is wasted as the camera rotates, so a second method involving continuous rotation at constant speed was developed. It is obviously important that the same concentration of radiopharmaceutical remains in the region being scanned throughout the procedure, thus speed is necessary for this reason as well as for patient comfort.

A number of dual-headed gamma cameras are now installed in nuclear medicine departments, (Fig. 5.8), and such systems can be used for the procurement of SPECT information with reduced acquisition time and improved sensitivity.

The mathematics of image reconstruction is complicated and beyond the scope of this volume. The reader is referred to (14), (15), for more detail. After processing, the images are analysed by a filtered back projection technique similar to that described in Chap. 2, to produce three dimensional images. The filter algorithm contains a correction for photon attenuation for each image projection.

The computer reconstruction allows SPECT studies to be presented on the screen either as a series of cross sectional slices or as a three dimensional display. Both have their place in diagnosis. Slices are more useful for some studies, while a 3-D display can be rotated continuously on the screen in order to examine organ defects.

SPECT Studies

SPECT is valuable in a number of areas already discussed under planar imaging, where extra information can be gathered from viewing body slices. Bone, liver and brain investigations are improved by adding SPECT images.

Brain Imaging with SPECT

Some newer pharmaceutical agents labelled with thallium-201 or iodine-123 (half life 13 hours), as well as technetium-99m are capable of crossing the blood-brain barrier and are retained in the brain for long enough to complete SPECT studies. They have important uses in the investigation of cerebral blood flow in stroke, transient ischaemic attack (TIA), and other diseases where blood flow is impaired.

Epilepsy is caused by abnormal electrical activity in the brain. Electro-encephalography (EEG) is valuable in the diagnosis of seizures, but SPECT can pinpoint the location of the focus of activity.

Degenerative diseases such as Alzheimer's disease show up with SPECT as areas of decreased activity in the parietal lobes and, less frequently, in the temporal, occipital and frontal lobes too (Fig. 5.9).

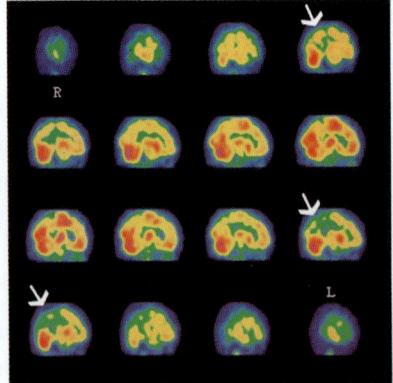 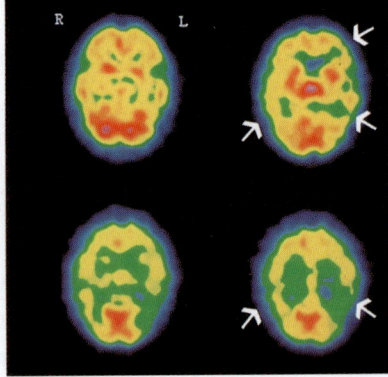

Fig. 5.9

Heart Imaging with SPECT

This technique is used extensively for studying blood flow (perfusion) in the muscles of the heart wall (myocardium). Slices in three different planes, the short axis and the long vertical and horizontal axes, with the heart at rest and when stressed demonstrate reversible, ischaemic problems and non-reversible defects or infarcts. Tc-99m based agents are commonly used, and the 'stressed' images are produced after exercise on a treadmill or by the injection of pharmacological agents which induce stress conditions without exercise, such as vasodilators.

Gated myocardial perfusion studies involve separating the various sections of the cardiac SPECT scans throughout the cycle of the heartbeat. This is a beneficial technique for investigating myocardial perfusion abnormalities which exist in conjunction with impaired ventricular function.

PET

Positron Emission Tomography (PET) has only been available in a few centres because it has been necessary to have immediate access to cyclotrons capable of producing positron emitting isotopes. Positrons are positive beta particles and the positron emitters used in scanning were all, until very recently, isotopes based on atoms found in the human body. Carbon-11, nitrogen-13, and oxygen-15, with half lives of 20.5 minutes, 9.9 minutes and 2.0 minutes respectively are incorporated into substances metabolised by the body such as glucose, water and carbon dioxide.

Isotopes with slightly longer half lives such as fluorine-18 (110 minutes) are also used. More recently, mini-cyclotrons have been designed specifically for use in nuclear medicine departments, and two generator-produced isotopes, gallium-68 (68 minutes) and rubidium-82 (76 sec) have been developed for PET studies. Rubidium is chemically similar to thallium.

The primary β^+ radiation that these nuclides emit penetrates only about 1 mm in tissue, but when a positron has slowed down to be nearly at rest, it is annihilated by a negative electron creating two photons of 511 keV which are emitted simultaneously in opposite directions. (If the positron is in motion at annihilation, the two quanta will not be exactly at 180°, but very nearly so.) By setting a coincidence requirement on pairs of opposing scintillation detectors measuring only at 511 keV, information about the position of the nuclide in the body can be obtained.

The technique as now undertaken was developed in St. Louis, Missouri by Michel Ter Pogossian and colleagues (16), but annihilation radiation had originally been used in nuclear medicine some thirty years earlier by J. R. Mallard and colleagues (17) at the Hammersmith Hospital, London, and by G. L. Brownell and colleagues in Boston, Massachusetts (18).

Mallard's group used arsenic-72 and arsenic-74 from the Hammersmith cyclotron in the detection of brain tumours. A pair of opposed scintillation detectors were moved across the patient's head to detect the annihilation radiation and the method, although crude and time consuming by today's standards, produced results which were more accurate clinically than information available from X-ray angiography at that time (19).

The PET Scanner

A PET scanner is composed of a large array of scintillation detectors with photomultiplier tubes which can be arranged around the body as in Fig. 5.10. Figure 5.10(a) shows how the slice width is controlled by slit collimation, while Fig. 5.10(b) shows a typical hexagonal array of detectors. Little or no collimation is required in this dimension because coincidence detection eliminates scatter.

Because of the high energy of the annihilation photons, only small corrections are required for tissue attenuation. The combined distance in tissue travelled by two photons is the same whether the annihilation takes place at, for example, either 'x' or 'y', so quantitative measurements can be made and the scanner can be calibrated to give absolute levels of radioactive concentration in tissue.

The scintillation detectors normally used in PET are made of bismuth germinate and are many times more sensitive than the scintillator in a conventional gamma camera. This extra sensitivity is necessary because of the relatively high energy of the radiation. Noise is very low so the effective

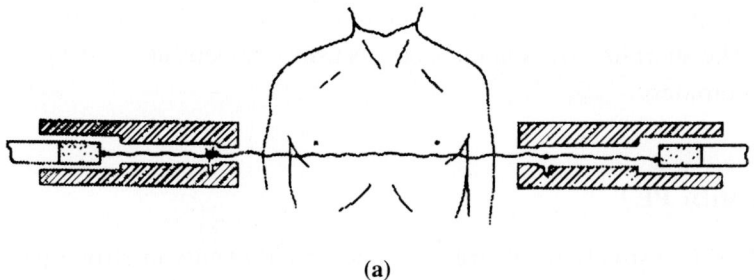

(a)

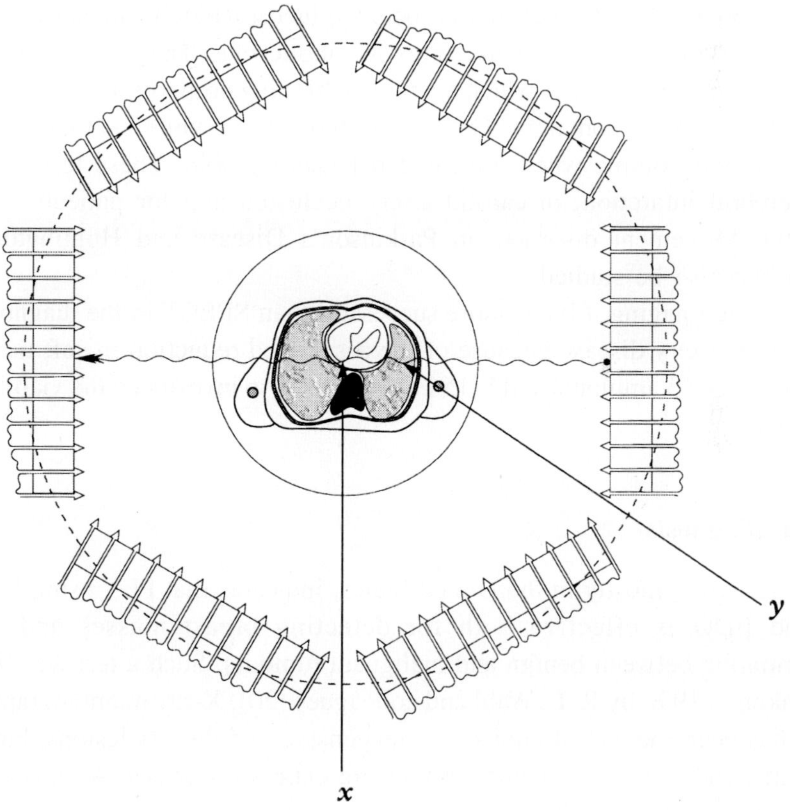

(b)

Fig. 5.10

dose to the patient can be approximately the same as for routine gamma imaging. The short half lives of the isotopes used compensate for the presence of the β^+ emitters.

Imaging with PET

For cerebral investigations, to track physiological events in stroke patients, PET using N-13 ammonia and O-15 labelled agents is used in preference to SPECT.

The uptake of F-18 labelled fluorodeoxyglucose (FDG) indicates local glucose utilisation rate which measures metabolic rate. The brain converts glucose into energy, so areas which are not functioning properly show lowered glucose levels. Water containing O-15 shows blood flow, while C-11 measures regional blood volume when used to label haemoglobin. These tests help after cerebral infarction, in carotid artery occlusion and for patients with dementia. Movement disorders in Parkinson's Disease and Huntington's Chorea can also be studied.

In cardiac imaging, PET is more successful than SPECT in the diagnosis of coronary artery disease because of the associated reduction in soft tissue attenuation. N-13 ammonia and F-15 FDG are used to investigate the viability of heart muscle.

Cancer Diagnosis

Recent research, mostly in the United States, indicates that PET using F-18 labelled FDG is effective both for detecting breast masses and for differentiating between benign and malignant tumours. Such a test was first undertaken in 1990 by R. L. Wahl and colleagues (20). X-ray mammography, which has been the test of choice, is inexpensive and detects lesions, but it gives little indication of whether lesions are cancerous or not. A dedicated PET scanner, much smaller than a whole body machine, has been designed and built by R. Freifelder and colleagues in Philadelphia (21). This has been found to be both accurate and cost effective and can be used also to monitor the progress of treatment.

Radioimmunoassay

The quantity of drug, hormone, enzyme, or disease-related antigen present in a sample of blood, body fluid or biopsy specimen can be investigated by the technique of radioimmunoassay. The principle of the technique involves an immune reaction between the substance of interest and specific antibodies to that substance. These tests are often carried out in departments of biochemistry or pathology rather than nuclear medicine.

When a specific antibody to a substance is mixed with the substance (S), a complex is formed which is chemically different from the constituents. If a known quantity, (s), of a radiolabelled analogue of the substance under investigation is added to the sample together with a quantity of unlabelled antibody insufficient to complex all the S and s, two forms of the complex are formed and the ratio of active to inactive parts allows the concentration of S to be measured.

The first reported work on radioimmunoassay came almost simultaneously in 1959, from the UK and the USA. R. P. Ekins at the Middlesex Hospital, London, used labelled thyroxine to estimate thyroxine hormone levels, while R. Yalow and colleagues in the USA reported on insulin assay using a similar technique. Yalow received the 1977 Nobel Prize in recognition of this major work.

The chief isotopes used are iodine-125 (half life 60 days, principle gamma energies 0.035 and 0.027 MeV), tritium (H-3), and Cobalt-57 (270 days, 0.122 MeV) and cobalt-58 (71 days, 0.81 MeV), used for tests to study vitamin B-12 absorption. The latter is called the Schilling Test and it helps to diagnose megaloblastic anaemia. The two isotopes can be counted simultaneously because of the different energies. This test has, in recent years, largely been replaced by biochemical analysis which does not require the use of radionuclides.

The activities involved are very small, of the order of kBq, and the counters used are often capable of handling samples in bulk. For very large numbers, the samples are moved sequentially over a sodium iodide crystal detector. For smaller numbers, measuring instruments containing 12, 16, or 20 individual crystals, one for each sample, are used.

Carbon-14 Urea Breath Test

Most patients with gastric ulcers have the bacterium *Helicobacter pylori* in their gut which triggers inflammation when it invades the mucous membranes. This bacterium has a high level of urease activity generating carbon dioxide. There are two possible isotopes which can provide diagnostic information of which the C-14 breath test is the most common.

The tests involve the ingestion of radioactive carbon which is converted by the bacteria into radioactive carbon dioxide which is exhaled by the patient. The presence of activity in the breath gives an indication of the amount of *H. pylori* present.

The effective dose to the patient with these procedures is about 0.1 mSv.

Activation Analysis

In a limited number of centres which have access to a reactor or particle accelerator, it is possible for activation analysis to be undertaken. This involves the irradiation of an inactive sample to induce radioactivity which is then assayed. A standard sample of material of known composition is irradiated under identical conditions and compared with the unknown sample to determine the composition. For accurate quantitative analysis, all atoms of each of the elements of interest must have the same chance of being activated. This means that there must be a uniform fluence of bombarding particles throughout the sample, and that the counting system must be independent of geometry.

For *in vitro* analysis, small samples of bone, tissue, etc. can be placed in high flux regions, for instance, the core of a reactor. By suitable choice of bombarding particles and energies, the induced radioactivity can be great and solid state detectors with good energy resolution can be employed to analyse the samples.

In vivo analysis must ensure that the radiation dose to the patient is acceptably low. The choice of bombarding particles is limited to photons or neutrons, as no others can penetrate the body to any reasonable depth. Photons are not much used because of the low values of photon activation cross section for most elements. Fast neutrons are attenuated significantly by the body, but if irradiation is carried out from more than one direction, neutron flux uniformity can be achieved.

From the 1970s, whole body counters were developed with high sensitivity and the ability to cope with patients of varying sizes (22).

Therapeutic Uses of Radionuclides

Radioactive isotopes are used to treat a variety of conditions. Some examples follow.

- Iodine is metabolised by the thyroid gland to produce thyroid hormones. An overactive thyroid gland can be suppressed by administering I-131 orally, in the chemical form of sodium iodide. Typically 200 to 400 MBq is given, some of which is excreted, but much of which is taken up by the thyroid for conversion into hormones. The gland receives a sufficiently high radiation dose, due to the beta emission of I-131, to reduce its metabolic function to the normal range.
- Some thyroid cancers can be treated in the same way, although with much higher doses. Not all thyroid tumours take up iodine, and those which do are usually treated by surgical removal initially. Up to 5 TBq of I-131 is administered in appropriate cases to destroy what remains and any secondary deposits elsewhere in the body. Patients are usually scanned on the gamma camera to locate these deposits and subsequently for follow up purposes.
- Phosphorus attaches itself to red blood cells. P-32, a beta emitter with a 14.3 day half life can be injected in the form of sodium phosphate to treat a condition in which the bone marrow is overactive and produces too many red blood cells, called polycythaemia rubra vera.
- Colloidal gold-198 and yttrium-90 are beta emitters used in the treatment of disseminated cancers in the peritoneal cavity and also for injection into joint cavities in the treatment of osteoarthritis. In the latter condition the walls of the synovial cavity are irradiated to reduce inflammation caused by a coating of unwanted material.

References

1. Hevesy G. and Chiewitz D., *Nature* **136**, 754–755, 1935.
2. Veall N. L., *British J. Radiol.* **23**, 527–534, 1950.
3. Mayneord W. V., *Some Applications of Nuclear Physics to Medicine*, British Institute of Radiology, Supplement 2, BIR London, 1950.
4. Herbert R. J. T., *Nucleonics* **10**, 37–39, 1952.
5. Curtis L. and Cassen B., *Nucleonics* **10**, 58–59, 1952.
6. Anger H. O., *Rev. Scientific Instr.* **29**, 27–33, 1958.
7. Bender M. A. and Blau M., *Progress in Medical Radioisotope Scanning*, New York, USAEC, 1962.
8. Datz F. L., *Handbook of Nuclear Medicine*, St. Louis, Mosby, 1993, ISBN 0-8016-7700-9.
9. Maisey M. N., Britton K. E. and Gilday B. C., *Clinical Nuclear Medicine* (2nd Ed.), New York, Chapman & Hall, 1991.
10. Mould R. F. (ed.), *Quality Control in Nuclear Medicine Instrumentation*, London, HPA Press, 1983, ISBN 0-90418-26-X.
11. Threefoot S. A. *et al.*, *J. Clinical Investigation* **28**, 661–670, 1949.
12. Kuhl D. E., chapter in *Medical Radioisotope Scanning*, vol. 1, pp 273–289, IAEA Vienna, 1964.
13. Bowley A. R. *et al.*, *British J. Radiol.* **46**, 262–271, 1973.
14. Kember N. F. (ed.), *Medical Radiation Detectors,* pp 143–153, Bristol IOPP,1994, ISBN 0-7503-0319-0.
15. Williams E. D. (ed.), *An Introduction to Emission Computed Tomography*, London IPSM, 1985, ISBN 0-904181-00-6.
16. Ter Pogossian M. M. *et al.*, *Semin. Nucl. Med.* **22**, 140–149, 1992.
17. Mallard J. R., Fowler J. F. and Sutton M., *British J. Radiol.* **34**, 562, 1961.
18. Aronow S., Brownell G. L. and Sweat W. H., *J. Nucl. Med.* **3**, 198, 1962.
19. Maisey M. N. and Jeffery P., *British J. Clin. Pract.* **45**, 265–273, 1992.
20. Wahl R. L., Cody R. L., Hutchins G. D. and Mudgett E. E., *Radiology* **179**, 765–770, 1991.
21. Freifelder R. and Karp J. S., *Phys. Med. Biol.* **42**, 2463–2480, 1997.
22. Boddy K. *et al.*, *Phys. Med. Biol.* **41**, 570–579, 1968.

Chapter 6

SOUND WAVES, MAGNETIC FIELDS, AND OTHER MODALITIES

Medical Ultrasonics

In contrast with the rapid development of the medical use of X-rays immediately following their discovery, ultrasound in medicine was very slow to be used. Sound waves with frequencies higher than those which can be heard by man had been recognised in the late 18th century on the realisation that bats navigated using ultrasound.

Sound is a form of energy which consists of mechanical vibrations. Unlike electromagnetic radiation, sound requires a medium through which to propagate in the form of waves. Several wave modes are possible, but diagnostic applications usually involve the use of longitudinal waves.

The particles in a medium vibrate backwards and forwards about their mean positions, so that energy is transferred through the medium in a direction parallel to that of the oscillations of the particles. The particles do not move through the medium, they simply vibrate to and fro.

Wavelength, λ, and frequency, f, of sound waves are related to the propagation velocity, c, by the equation:

$$c = f \lambda$$

For example, in water, where c is approximately 1500 metres per second, sound waves with frequency 1 MHz (10^6 hertz) have wavelength 0.15 cm. Unlike light, the velocity of sound in matter is practically independent of frequency. Velocity depends on the medium through which the sound travels; the greater the density, the lower the velocity.

151

152 Light, Visible and Invisible

The lower frequency limit of the ultrasonic spectrum is generally taken to be about 20 kHz, although most diagnostic applications employ frequencies in the range 1–15 MHz.

Like electromagnetic radiation, sound undergoes reflection and refraction at an interface between two different media. It is the reflections, or echoes, from different tissues which produce ultrasound images. Diagnostic ultrasound has special uses in the imaging of soft tissues whose densities are too similar to produce contrast on X-ray films. In obstetric imaging where early information about foetal development is important, ultrasound is considered very much less hazardous than X-rays.

Historical Development

The scientific study of sound waves was begun in the mid 19th century. In 1877 and 78, Baron Rayleigh, (John Strutt), published two volumes entitled, "The Theory of Sound". Rayleigh was James Clerk Maxwell's successor as Cavendish professor of physics at Cambridge and in a position to encourage scientists to explore this area of physics. Rayleigh was physics Nobel prize-winner in 1904.

The foundations for the generation and detection of ultrasound were laid by the Curie brothers, Pierre and Jacques, when they discovered the piezoelectric effect in Paris in 1880. They found that when mechanical pressure was applied to some materials such as quartz and certain ceramics, an electric charge was produced. They subsequently found that the reverse was also true, that an oscillating potential applied across a quartz crystal caused it to expand and contract alternately. This produced vibrations which were transmitted as sound waves. This is the principle of the modern transducer — a term coined to describe a device which provides a coupling between electrical and mechanical energy.

The *Titanic* disaster of 1912 was followed by efforts to detect undersea obstacles by bouncing sound waves off them. The earliest echo sounding machine was developed in Paris by Paul Langevin and colleagues and was used by the French in World War I to detect enemy submarines. By the time of the second World War, detection apparatus had been greatly improved and developed with the addition of advanced electronics to create the system called SONAR, (sound navigation and ranging) used by the British, the Americans and the Japanese.

Between the wars, the destructive effects of ultrasound were applied in medicine and in industry, following Langevin's observation that ultrasound beams could destroy small sea fish. In Germany, ultrasound was applied in efforts to destroy cancerous cells, while elsewhere in Europe and America it was used in physiotherapy as a treatment for muscular strains.

In the early 1950s, it was discovered that a ferroelectric material, lead zirconate titanate, had piezoelectric properties far superior to those of quartz. This led to much improved sensitivity and pulse performance.

Diagnostic Methods

Tranducers

Most applications of diagnostic ultrasound require a narrow beam. This is usually generated by a disc of piezoelectric material electrically excited by means of two thin metal electrodes on opposite parallel surfaces. When alternating voltage is applied between the electrodes, the piezoelectric effect causes a synchronous variation in the thickness of the transducer. It can be made to emit sound of any frequency by driving it at that frequency, but the largest output is produced when the frequency at which the crystal vibrates produces a wavelength in the transducer equal to twice the thickness of the piezoelectric disc. This is because the front face of the transducer emits sound both backwards and forwards, and the backward wave is reflected forwards from the back face. Reinforcement, i.e. constructive interference, occurs if this reflected wave is in step with the forward emission from the front face.

The rear electrode is live and the front one is connected to the metal case which is earthed. A schematic diagram of a transducer probe is shown in Fig. 6.1. The matching layer consists of a thin plastic disc which serves the dual purpose of protecting the transducer, and as an aid to maximum transmission, which occurs when its thickness is one quarter of the wavelength. The backing block absorbs backward travelling waves produced by the back face of the vibrating disc. Further information on ultrasonic transducers can be found in Refs. (1), and (2).

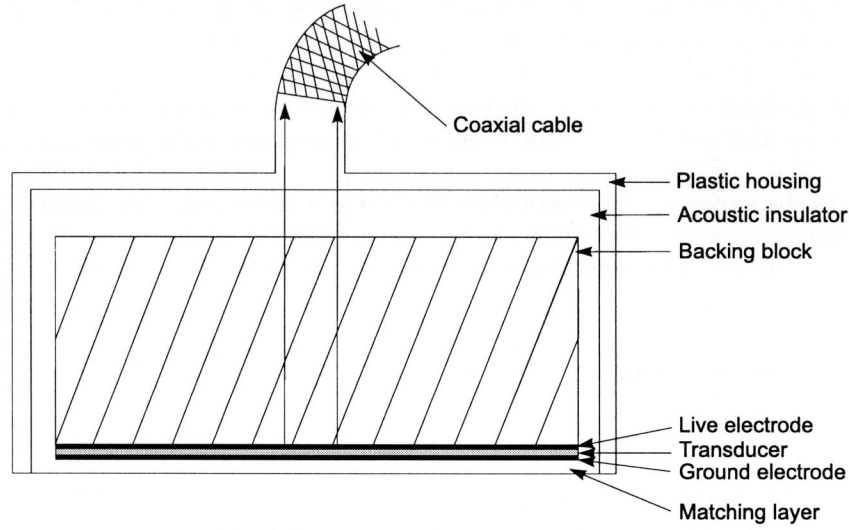

Fig. 6.1. Diagram of a transducer probe.

Most ultrasonic diagnosis is based on the reflection of ultrasonic waves at boundaries between different tissues in the body. At a boundary, some energy is reflected while some passes through the barrier and may be reflected at deeper boundaries. Because some ultrasound is reflected at the boundary between the outside air and the patient's skin, coupling oil or gel is applied between the probe and the patient to keep reflection to a minimum. The maximum penetration in tissue is limited by the attenuation of the ultrasound in passing through the body.

In many applications the probe acts as both transmitter and receiver.

A-Mode

Amplitude mode, or A-mode, scanning was the first to be developed in medicine. A short duration pulse is emitted from a probe which is connected to a cathode ray tube in such a way that the screen registers a vertical spike when a pulse is emitted. The screen registers a second spike when the transducer receives a reflected pulse. The time between the two spikes is the time taken for the pulse to travel to the tissue boundary and back and is proportional to distance, while the height of the spike is proportional to the strength of the echo.

The first medical application of this technique was by the Austrian brothers Karl and Frederick Dussik (one a physician and the other a physicist) in 1947, when they tried to demonstrate boundaries in the brain (3). In the United States in 1949, George Ludwig showed the presence of gall stones using apparatus adapted from naval military research (4).

A-mode scanning became fairly widely available in medicine in the 1960s.

B-Mode

Brightness mode, or B-Mode, scanning gives two dimensional pictures. A-scan information is converted into dots of variable brightness corresponding to the strength of the returning echo. Each scan line of dots is coupled with a time-base and with equipment which links the direction and position of the B-scope time-base on a cathode ray tube to those of the ultrasonic beam on the patient. Thus a two dimensional picture can be built up as shown in Fig. 6.2.

John Wild, an Englishman working at the University of Minnesota in 1952, started real time scanning with an oscillating transducer (5). A standing 'patient' was immersed up to his chin in a water bath and the transducer moved to and fro around his neck. This led to the development of the first contact scanner in 1956 by Ian Donald in Glasgow, Scotland, working with engineers from the company Kelvin and Hughes. This avoided the need for a water bath but allowed free movement of the transducer in one plane. The first scans of foetuses were produced with this apparatus.

B-mode is also known as 2-D or Grey Scale imaging, and in modern B-scanners the image is automatically scanned in a series of frames fast enough to demonstrate the motion of tissues.

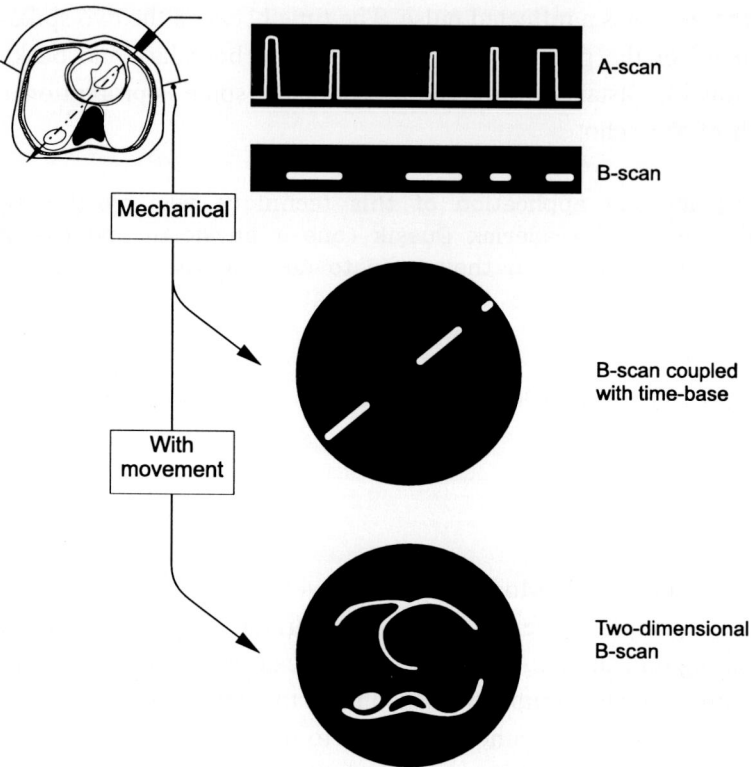

Fig. 6.2. The basic principles of two-dimensional ultrasonic scanning.

Images were always black and white until grey scaling was introduced in the mid 1970s by using a digital scan converter. Digital echo signals are read electronically and the signals used to drive the electron gun of a TV monitor. With this method, strong echoes appear as lighter shades of grey than weaker ones.

M-Mode

Where rapid tissue movement takes place, as with the beating heart, M-mode scanning is used. The technique allows a B-mode image to be frozen on the screen. A beam from a stationary transducer is then directed along the line of interest, intersecting the rapidly moving surface as nearly as possible at right angles. Echoes can be displayed on the screen as a line of moving bright dots. Thus, the variation of a single static B-mode beam line with time is shown. This can be recorded and played back as required.

Doppler

The Doppler effect is a change in the frequency of sound which is reflected from a moving surface. Movement of the surface towards the source results in compression of the wavelength of a reflected wave, and movement away causes extension of the wavelength. Since velocity of propagation is constant, these changes in wavelength produce corresponding changes in frequency.

At normal incidence, the Doppler shift frequency, f_D, is

$$2vfc^{-1}$$

where v is the velocity of the reflecting boundary towards the source, f is the frequency of the incident wave and c the propagation velocity.

For oblique incidence at angle I,

$$f_D = 2vf \, (\cos I) \, c^{-1} \, .$$

Thus the maximum Doppler shift occurs with normal incidence.

The Doppler effect is named after Johann Doppler, an Austrian physicist, who investigated it in 1842. The earliest medical use of Doppler was probably by Shigeo Satomura in Japan in the early 1950s (6).

Doppler is used to provide information about the velocity of moving structures and particularly to study blood flow. The probe uses two transducers placed at a slight angle to each other, one as transmitter and the other as receiver. Frequencies in the range 2–10 MHz are chosen according to the depth of the vessel under examination. The Doppler beat signal can be displayed on a screen or heard on headphones. The latter is often the method of choice because the human ear is very sensitive to changing sound patterns.

It is impossible to locate the exact position of the moving reflector with continuous wave Doppler, or to distinguish between the flow in overlapping vessels. Continuous wave Doppler forms the basis of the ultrasonic stethoscope used to monitor foetal heartbeat.

In duplex scanning, real-time B-scans are combined with Doppler to give information on position and diameters of vessels. This allows volume flow rates to be calculated. The addition of colour is very useful and colour Doppler shows both velocities and variations in direction and velocity of flow. It is particularly useful in detecting regurgitant jets and areas of turbulent flow within vessels.

Recent developments include power Doppler, where the mapping of the strength or amplitude of the Doppler signal rather than the velocity and direction, overcomes some of the problems of conventional colour Doppler. It can give a more sensitive map of perfusion, although at the expense of directional information. This is also known as Energy Mode, Angio Doppler, and Colour Angio.

Tissue Doppler Imaging, allows low frequency, high amplitude signals from moving structures, which might be seen as artefacts and so filtered out on conventional Doppler studies, to be shown as a colour map over the B-mode image. This can be of value in cardiac studies.

Three-Dimensional Imaging

Real time 3-D imaging is not available, but some scanners can now show pictures of volumes of tissue. The methods and capabilities of different systems are variable, and collection times are reducing.

Diagnostic Investigations

The Heart and Blood Vessels

This topic has largely been covered in the text already. M-mode imaging is a very useful technique for the study of the beating heart, while Doppler is invaluable in all blood flow studies.

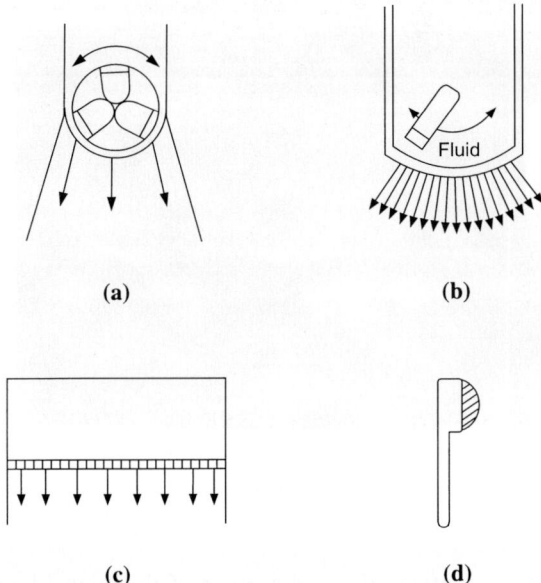

Fluid

(a)   (b)

(c)   (d)

Fig. 6.3. Types of transducer: (a) Mechanical rotating. (b) Mechanical oscillating. (c) Electronic linear array. (d) Endoluminal.

The Brain

Leskell introduced ultrasonic encephalography in Scandinavia in 1954 (7), in an endeavour to show the position of the midline of the brain following head injury. This was a simple A-mode technique which was subsequently extended to produce images of brain tumours, haematomas, and other subdural and epidural abnormalities.

More recent applications include the use of Doppler to investigate carotid artery blood flow in atheroma, or narrowing of the arteries, through disease.

Obstetrics and Gynaecology

This is an extremely useful area of diagnostic ultrasound. In pregnancy, where exposure to ionising radiation is not recommended, B-mode scanning

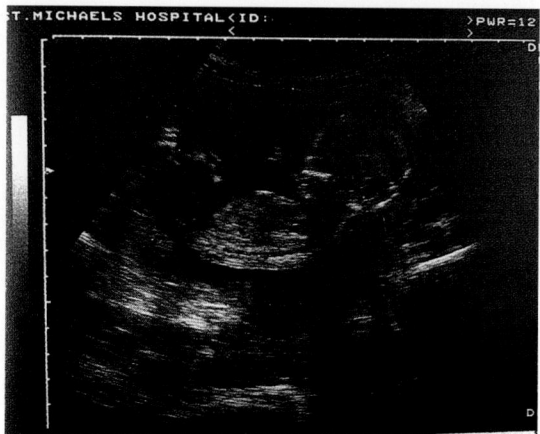

Fig. 6.4

gives an accurate estimation of the diameter of the foetal head and outlines of organs. Fig. 6.4, is an ultrasound scan of a four month old foetus. Foetal heartbeat is measured using A-mode or Doppler.

Gynaecological tumours normally displace bowel and often lie in close proximity to the abdominal wall making them easily accessible to ultrasonic scanning.

Ophthalmology

B-scanning is of great benefit in producing cross sectional images of the eye and demonstrating the constituency of visible lesions within the eye. When the cornea is opaque or the lens clouded with cataract, ultrasound may be the only useful diagnostic method. The technique is also used to diagnose posterior retinal detachment and retinal tears.

It is routinely used to demonstrate the growth of the eyes in children with developmental disorders, and to help diagnose problems elsewhere in the orbit.

Ischaemic eye disease, in which there is reduced blood flow in the retina, is investigated with Doppler.

The Liver

Liver cysts and abscesses can be well demonstrated with ultrasound. It is also valuable for the diagnosis of cirrhosis, multiple secondary tumours, and polycystic liver disease.

For more detail on ultrasonic diagnosis the reader is referred to Refs. (8)–(10).

Bone Densitometry

The evaluation of bone density by ultrasound was first reported in 1984 by C. M. Langton and colleagues (11), and it was used to assess osteoporotic bone soon afterwards.

Two parameters can be measured: Broadband Ultrasonic Attenuation (BUA), and Speed of Sound (SOS). Osteoporotic bone is less dense than normal bone and so attenuates ultrasound to a lesser degree. Normal bone has an intact and continuous trabecular architecture and the speed of sound is greater than in the architecturally degraded bone of osteoporosis. The Quantitative Ultrasound Index (QUI) is an algorithm combining BUA and SOS and is sometimes referred to as a measure of bone stiffness.

There are difficulties associated with obtaining accurate measurements because of the presence of overlying tissue and of poor contact with the transducers, but results from ultrasound studies are now sufficiently advanced to demonstrate osteoporosis and predict its onset. The technique is simple, quick and cheap, usually involving measurements on the heel bone. It is also attractive because it does not involve ionising radiation.

Ultrasound in Physiotherapy

The death of fish which swam into the ultrasonic fields of early SONAR, and the painful sensation felt by placing the human hand in the beam, led to the realisation that there might be surgical and therapeutic applications of ultrasound.

Different tissues absorb ultrasound differently. At 1 MHz, soft tissues absorb at the rate of 1 dB per centimetre, while muscle absorbs at 2 dB per centimetre. Thus, a thickness of less than 3 cm of tissue will result in half of the energy being absorbed. Attenuation increases with frequency, so that at 3 MHz it is three times this value, so higher frequencies produce greater effects in superficial tissues.

Ultrasound was first used in physiotherapy in the 1950s with the object of producing tissue heating. Heating causes increased blood flow and capillary permeability, fibrous tissue becomes more extensible and muscular relaxation occurs.

Non-thermal effects also appear to be important and this is why pulsed apparatus is often used.

At high field strengths there is a danger of cavitation occurring (see below).

Hazards

Diagnostic ultrasound is considered to be completely safe, and there is no confirmed evidence of damage from ultrasound imaging.

The critical organ for airborne ultrasound is the ear, and reported effects of ultrasound include temporary threshold shifts in sound perception, fatigue, headaches, nausea and tinnitus. However, since airborne ultrasound is usually accompanied by simultaneous high exposure to audible frequencies, it is not possible to attribute these effects purely to ultrasound.

At high levels, there are dangers of:

1. Local heating leading to cell damage, but heat can often be dissipated through blood flow.
2. Acoustic streaming when ultrasound is scattered in inhomogeneous tissues. In such a situation, the different components vibrate with different amplitudes and can move relatively to each other causing tissue damage.
3. Cavitation in which small bubbles, present in all normal liquids, grow in size under the action of mechanical vibration. When they reach a certain size in relation to the ultrasonic wavelength, (about 6 mm diameter at 1 MHz), they behave as resonant cavities and their vibration amplitudes become very large.

4. Mechanical damage to cell membranes due to violent acceleration.

Safety Testing

Regular checking must include the measurement of power output of transducers by 'weighing' the sound pressure with a force balance or by measuring heat produced using a calorimeter. Time-averaged intensity should not exceed the safe level of 100 mW cm^{-2}, and total sound energy should not exceed 50 J cm^{-2}.

Resolution is checked using a phantom containing parallel wires in a perspex block; while sensitivity, dynamic range and distance accuracy require a phantom in which reflections can take place at several levels.

Grey scale performance and Doppler functions are checked with more complex phantoms (12).

Magnetic Resonance Imaging

Magnetism has been recognised since ancient times, and magnetic compasses have been in use since the Middle Ages. Terrestrial magnetism, the fact that the earth behaves like a giant magnet, was first described by William Gilbert, physician to Queen Elizabeth I, in 1600.

In 1820, it was demonstrated by Oersted in Denmark, that an electric current could produce a magnetic field. Electricity and magnetism were definingly connected by Maxwell's equations in 1865.

In 1924, Wolfgang Pauli, an Austrian physicist working, at the time, on atomic structure in the University of Hamburg, Germany, suggested that certain atomic nuclei behaved like tiny bar magnets by virtue of their spin and associated electrical charge. Felix Bloch, a physicist from Zürich who had emigrated to the United States in 1934, worked on this and associated ideas. In 1946, he published papers on the detection of signals from atoms with odd numbers of sub-atomic particles which were placed in strong magnetic fields, and where the spin orientation had been disturbed by brief pulses of radiofrequency radiation. Edward Purcell, at Massachusetts Institute of Technology, produced similar findings at about the same time. This established the basic principles of nuclear magnetic resonance.

Bloch and Purcell both undertook experiments on human subjects. A strong proton NMR signal resulted when a finger was inserted into a radiofrequency coil, and subsequently signals were obtained from the human head (13). However since very uniform magnetic fields are required for accuracy, magnet technology was not sufficiently advanced for worthwhile results to be produced. Bloch and Purcell were awarded the 1952 Nobel Prize for Physics for their discoveries.

The fascinating summary of the history of NMR and its clinical application was provided by E. R. Andrew in 1984 (14).

Scientific Principles

Atomic nuclei are made up of protons and neutrons. Protons have spin and thus electrical charge. They move about within the nucleus and, since a moving electrical charge is an electric current, it is accompanied by a magnetic field. Hence, a proton has its own magnetic field and can be considered as a small bar magnet.

Paired protons spin in opposite directions thus cancelling the magnetic moments. Unpaired protons, as in nuclei containing an odd number of protons, produce a magnetic moment for that particular nucleus. The easiest example is hydrogen whose nucleus consists of a single proton and no neutrons.

Normally, protons are aligned in random directions as shown in Fig. 6.5(a). When a strong magnetic field is applied, the protons realign themselves with their axes either parallel or anti-parallel to the magnetic field. This is usually referred to as 'spin up' and 'spin down' respectively. There is a small excess in the parallel direction since this state needs less energy, Fig. 6.5(b). The protons continue to spin on their axes but the static magnetic field causes them to 'wobble' in a regular manner called precession, as shown in Fig. 6.6(a). The spin axis tilts and rotates around the direction of the magnetic field with a fixed frequency called the Larmor frequency, (after Sir Joseph Larmor, the Irish born physicist who spent 50 years in the University of Cambridge, UK from 1885 and who made many discoveries in the area of sub-atomic particles). The stronger the magnetic field the faster the proton precesses. The frequency of precession (ω) is proportional to the product of the field strength (B) and

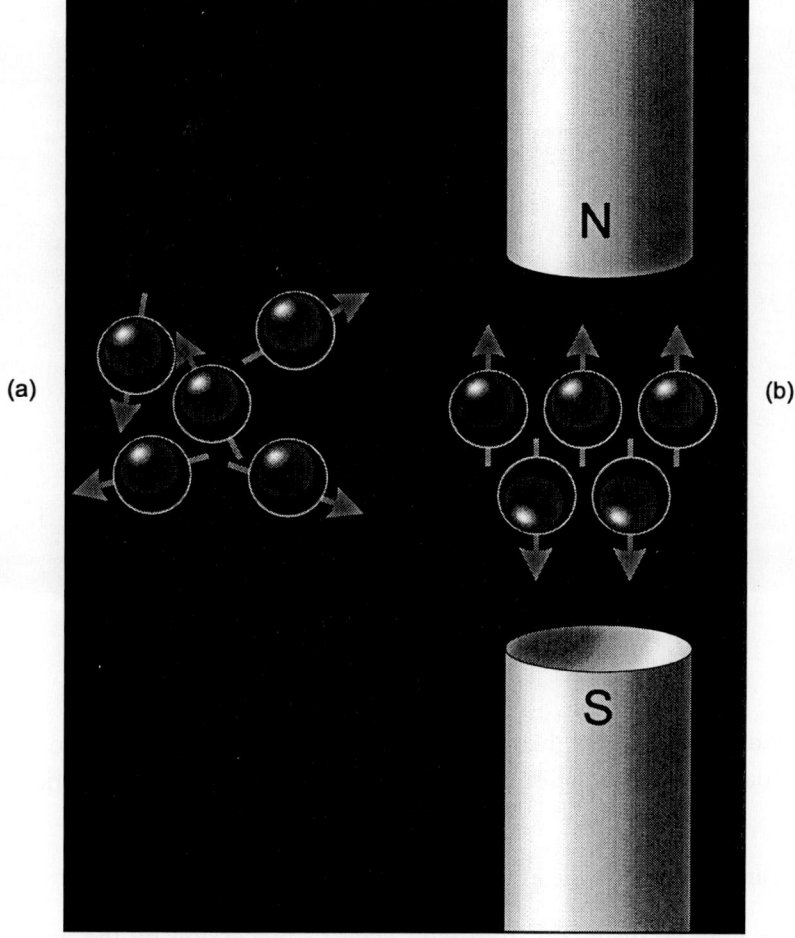

(a)   (b)

Fig. 6.5

the property of the nucleus called the gyromagnetic ratio (γ). So, the Larmor equation is:

$$\omega = \gamma B$$

The value of the gyromagnetic ratio for hydrogen nuclei is 42.6 MHz per tesla of magnetic field.

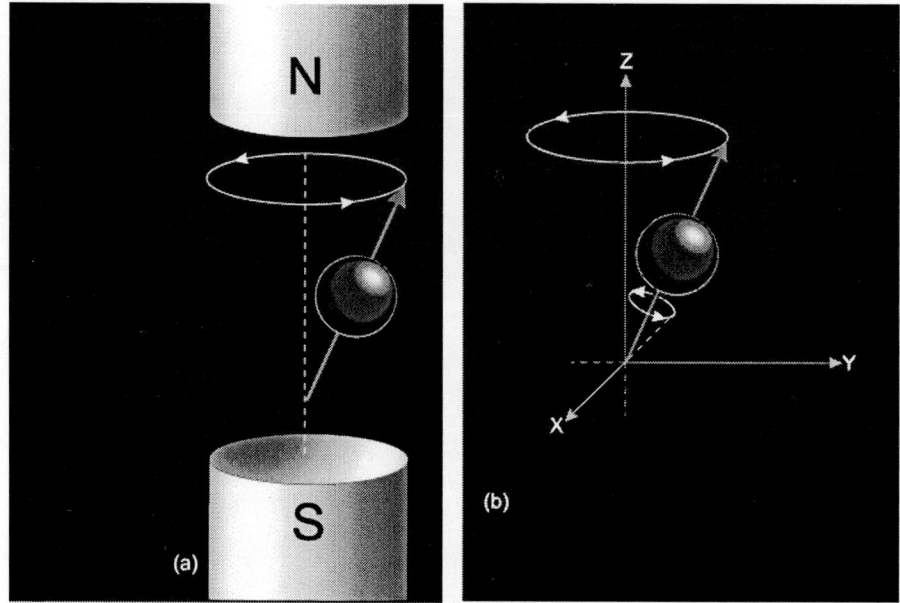

Fig. 6.6

Figure 6.6(b) shows the normal field co-ordinate system used to describe the situation with direction z representing the applied magnetic field.

Resonance

Protons may be disturbed from their state of precession by the application of an alternating magnetic field perpendicular to the main field. This is usually provided from an RF generator.

This causes two effects:

1. Some of the 'spin-up' protons pick up energy and turn 'spin-down'.
2. The protons are pulled into synchronism making them precess in step (or in phase).

The first results in a decreasing magnetic field along the z axis, and the second establishes a new magnetisation vector in the x-y plane.

Relaxation

When the radiofrequency pulse is switched off, the protons return to their former state. The magnetisation in the z direction increases again and this longitudinal relaxation is described by a time constant T_1. Additionally, the transverse magnetisation decreases and disappears, and this transverse relaxation is described by time constant T_2. The relaxation times are different with T_1 longer than T_2. The measurements of T_1 and T_2 for hydrogen nuclei, each of which consists of a single proton, provide the images of MRI. Hydrogen is very abundant in the human body which is made up largely of water. Table 6.1, gives a list of typical values, in milliseconds for parts of the body.

For a fuller explanation of the physics of MRI the reader is referred to specialist textbooks including (15)–(17).

The first machines operated purely on differences in T_1, which was found to be higher in malignant than in normal tissue, and it was thought of as a possible new diagnostic test for cancer. The differences in T_1 between the material of body organs was also found to be significant which extended the possible range of use of the technique.

The whole body scanner built at Aberdeen by J. R. Mallard and colleagues (Fig. 6.7) was the first, and was based upon a 0.04 T standing magnetic field. They were closely followed by P. Mansfield's three teams at Nottingham who employed slightly different techniques.

Table 6.1

Material	T_1	T_2
Body Fat	250	80
White Matter	650	90
Grey Matter	800	100
Liver	400	40
Kidney	550	60
Spleen	400	60
C. S. F.	2000	150
Water	3000	3000

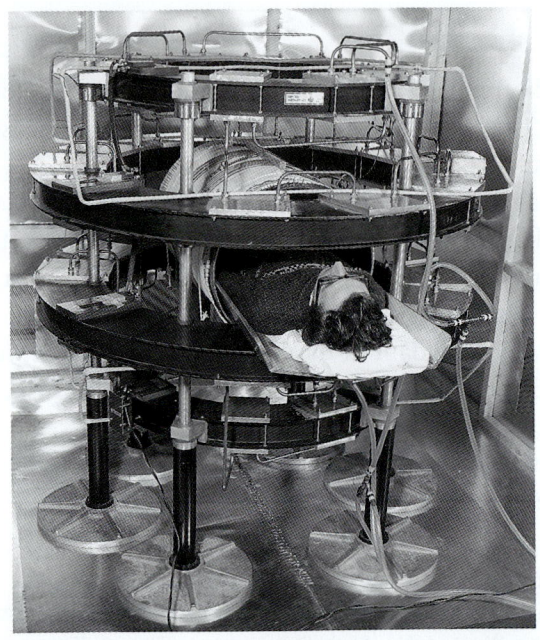

Fig. 6.7

Three different types of magnet units were developed. Permanent magnets made of ferromagnetic material were extremely heavy; typically, a 0.3 T magnet weighed 100 tons. Permanent magnets made of rare earth alloys produced higher fields with less weight. Some permanent magnet systems are still used, usually for specific tasks such as head, neck and extremity investigations which do not require high field strengths. Resistive magnets which generate magnetic fields by current flow within coils of wire proved more satisfactory but generated a lot of heat.

By 1980, the Oxford Instrument Company had perfected the development of supercooled, superconducting magnets with field strengths up to 2 T, but in somewhat restricting solenoids. Eventually, magnets with 100 cm diameter bores became available so that every part of the body could be investigated. Cooling is provided using both liquid nitrogen and liquid helium.

Further developments of the technique to produce the sophisticated MRI scanners now familiar in hospitals were undertaken chiefly in the United States, Japan, and Germany where far greater financial resources were available.

Figure 6.8, shows the arrangement of coils in a modern MRI machine. The outermost coils carry DC which produces the very uniform and strong magnetic

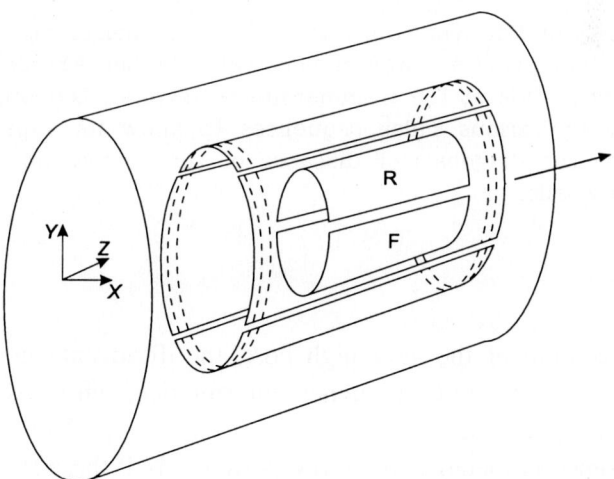

Fig. 6.8. Arrangement of the coils in an MRI machine.

field in the z direction. The patient lies inside the solenoid in the prone or supine position. Surrounding and close to the patient, the innermost coils are connected to an RF generator which produces the rapidly alternating magnetic field in the x-y plane.

MRI for Diagnosis

Magnetic Resonance Imaging does not use ionising radiation and is therefore considered to be a safe modality. Having been in regular use for at least fifteen years, no adverse effects have been reported.

There are significant diagnostic advantages too. MRI scores over conventional X-rays and X-ray CT in that the computer reconstruction allows for changes in T_1 and T_2 weighting which can change the contrast within an image and thus emphasise different structures. MR can provide transverse, coronal or sagittal slices, or indeed slices in any orientation, from one set of measured data. It is possible to visualise vascular structures without the use of the sort of contrast media used in radiography for angiography etc.

The main advantage over ultrasound is that overlying bony structures do not cause degradation of the image.

The superiority of MR over both ultrasound and nuclear medicine studies for imaging liver disorders was demonstrated by the Aberdeen group as early as 1980, while at the Hammersmith Hospital, London, Young and colleagues used various pulse sequences to show its superiority over X-ray CT for the diagnosis of diseases of the central nervous system shortly afterwards.

Problems

With the exception of the very high cost, the disadvantages of MRI are physical. Background radiofrequency interference, which is particularly prevalent in hospitals and clinics, makes installation problematical as does the near presence of metallic structures, particularly if these structures move about. Additionally, the MRI unit produces a magnetic field elsewhere than within the unit itself, and this can affect the operation of nearby medical equipment and computer discs. Shielding can be, and is, provided by building a Faraday cage around the MRI equipment.

Relatively small metallic objects such as scissors and surgical tools, and larger items like oxygen cylinders, are very dangerous if brought into the vicinity of an MRI scanner since they are rapidly drawn into the imaging cavity.

The time taken for imaging is much longer than the corresponding time required to produce radiographs and ultrasonic scans. It is often also longer than for nuclear medicine studies, although manufacturers are developing faster systems.

Clinical Problems

The construction of the machine makes it inherently claustrophobic as well as presenting difficulty in monitoring patients during investigation. It is also noisy because of the repeated switching of the gradient fields, which can distress some patients. The rapidly changing magnetic fields may induce electric currents within the body. In the heart there is a risk of ventricular fibrillation at high field strengths, so MRI is contraindicated in patients with severe cardiac disease. Radiofrequency fields can cause tissue heating. This is usually dispersed by blood flow, but might be a problem in the lens of the eye and other areas with little flowing blood.

Implanted metallic objects present problems. Surgical clips may move within the patient and cardiac pacemakers and other implantable devices containing metals can be affected. It is necessary to remove hair grips, ear rings and other metallic ornaments.

Non-metallic implants such as hip prostheses may become hot, because of the RF fields, causing severe discomfort to patients.

In the UK, the National Radiological Protection Board has produced guidance on the safe use of Clinical Magnetic Resonance (18). Other countries also issue guidance.

With due regard to the clinical difficulties described above, MRI can be used for the complete range of diagnostic investigations either on its own or in conjunction with other techniques. MRI is the modality of choice for imaging the central nervous system and muscles and tendons around joints. Tumours are well visualised. Figure 6.9 is an MRI scan of the knee joint.

Functional MRI

Magnetic Resonance angiograms can be generated in which flowing blood appears brighter than stationary tissues. This allows the possibility of such techniques as 'brain mapping' to establish which areas of the brain are responsible for particular functions like speech or movement by noting areas of increased blood flow. If the fingers are moved, for instance, there is activity somewhere in the brain which causes an increased blood supply to the nerve cells in that area.

Contrast Agents

Intravenous contrast agents have been produced. Agents are often based on gadolinium which was first used by Runge and colleagues both orally and intravenously in the early 1980s (19). Gadolinium is paramagnetic,

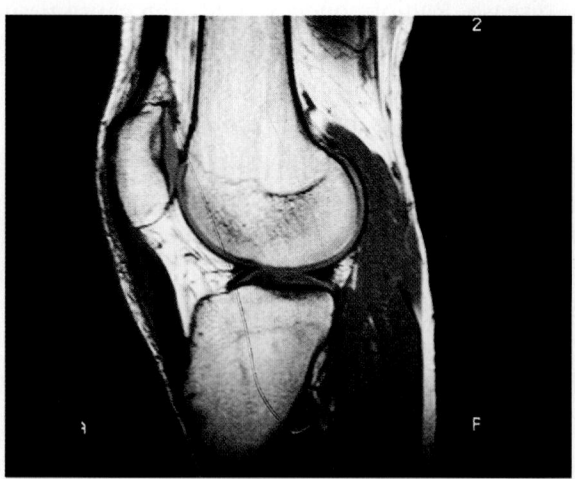

Fig. 6.9

and paramagnetic materials have the property that some of the orbiting electrons are not paired. Their magnetic fields do not cancel each other and their net magnetism is as much as a thousand times greater than nuclear magnetism.

Gd has seven unpaired electrons so is strongly paramagnetic, chemically it is soluble in water when chelated with diethylenetriamine pentaacetic acid (DTPA). Although it is not itself visible on the MR image, it affects a large number of hydrogen nuclei close to it so that the relaxation times are shortened, the effect being greater in T_1 than T_2.

Gd-DTPA is able to cross the blood-brain barrier in conditions such as infection, ischaemia and malignant tumours, thus improving the accuracy of brain imaging.

Tumours of the breast can be seen with MRI imaging, and diagnosis can be greatly improved with prior intravenous injection of Gd-DTPA. Contrast enhanced MRI is the investigation of choice in women with dense breasts or breast implants. Mathematical modelling has demonstrated significant differences between malignant and non-malignant lesions (20).

Magnetic Resonance Spectroscopy (MRS)

This technique involves magnetic resonance of other elements in addition to hydrogen. It is used to identify various chemical states without destruction of the sample, to give information about metabolism of chemical compounds.

Phosphorus-31 is, at present, the most useful element in use, but studies of drug metabolism using fluorine-19 are under investigation.

Because of chemical shift, phosphorus nuclei have different resonant frequencies when bound in inorganic salts. *In vivo*, P-31 spectra have three peaks due to adenosine triphosphate, (ATP), which is a major source of cellular energy, phosphocreatine, a short term energy reserve, and inorganic phosphate, a product of the breakdown of ATP.

In a high magnetic field and using a broad band RF pulse, all of these can be made to resonate. The magnetic resonance signals from a given volume of tissue can be analysed as a frequency spectrum and each of the salts can be imaged separately to allow the study of metabolism.

A field strength of 2 T or more is required to give sufficient signal strength, so the technique can only be performed with the use of MRI units which have superconducting magnets.

The metabolism of tumours differs from that of normal tissues. Phosphorus-31 MRS provides diagnostic information on tumour type and grading. It can also be used to monitor the efficacy of cancer treatment. Tumours studied include sarcoma, non-Hodgkins lymphoma and breast cancers (21). Hydrogen MRS has been used to study brain tumours and prostate cancer (22).

This is still very much a developing technique but it may be that magnetic resonance spectroscopy will become more important than magnetic resonance anatomical imaging.

Quality Assurance

The regular QA programme for an MRI scanner must include checks on the homogeneity of the magnetic field. Special probes have been developed for this measurement. Contrast, resolution, and image distortion are checked with suitable test objects. The RF parameters must also be measured.

The Future

Elsewhere than in the medical field, physicists are developing MR to exploit a whole range of applications. For instance, it is now possible to examine the distribution of different components in material with a resolution approaching 1 mm (23). The medical implications of these investigations are easily imagined.

Diagnosis with Infra-Red Radiation

Scientists have been shining various wavelengths at patients to try to discover disease and abnormalities for a very long time. In the 1970s it was discovered that the absorption of certain wavelengths in the near infra-red region was dependent upon the state of oxygenation of haemoglobin (24). Thus, a technique called NIR spectroscopy was developed which is now in routine use to monitor blood supply and tissue metabolism. There are two particular uses which are important in the clinical situation. The first is the monitoring of newborn premature babies, while the second is used routinely to provide information on anaesthetised patients or those in intensive care whose oxygen supply is being delivered artificially.

The Pulse Oximeter

As its title implies, this device measures both pulse rate and blood oxygenation. It is usually attached to a finger, but toes and ear lobes can also be monitored.

The finger is inserted into a small probe, like that shown in Fig. 6.10. Wavelengths generated by red (660 nm) and near infra-red (940) nm light-emitting diodes pass through the finger to a silicon photodetector which measures the relative absorption of the two wavelengths to measure the 'redness' of the blood. Oxygenated haemoglobin and reduced haemoglobin, which does not contain bound oxygen, exhibit different absorption characteristics at the two wavelengths.

Blood passes through the finger in pulses so the heartbeat can be measured incidentally.

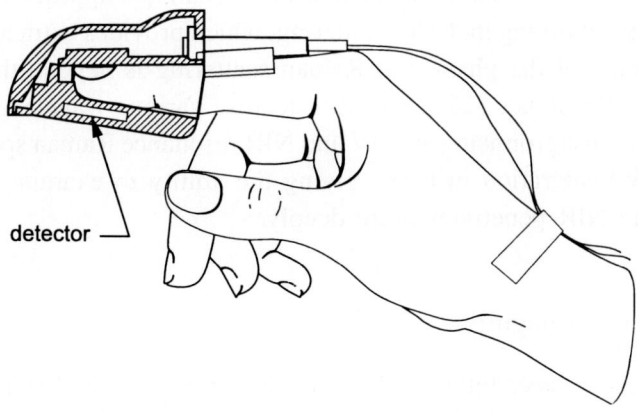

Fig. 6.10. Pulse oximeter. Finger probe attachement.

Diagnosis with Visible Light

Optical Spectroscopy

Biomedical optics has been in use since the mid 19th century when stains and dyes were first used to provide optical differentiation of cellular structures on microscope slides.

The recent development of tiny light sources such as lasers, and of small detectors and fibre optic probes, has opened a new chapter in which light is shone onto tissue and the reflected signal analysed. Light is scattered from tissue but the re-emitted light is often influenced by biochemical and physiological changes in the sample. This leads to the earlier detection of abnormalities, particularly of malignant and pre-malignant conditions, than by visible inspection, and the method is also non-invasive. Several different optical interactions are in use and being researched as diagnostic methods (25).

Laser induced fluorescence spectroscopy looks at fluorescent emissions after excitation with laser light. Major differences are noticed in the fluorescent emission spectra of normal and malignant epithelial tissue.

Elastic scattering spectroscopy investigates light scattered from tissue irradiated with white light or selected wavelengths by constructing intensity spectra.

The newest, and probably the most useful, technique appears to be Raman spectroscopy, involving inelastic scattering, which provides extra information. An explanation of the physics of Raman scattering is beyond the scope of this volume. Reference (26) gives full details. Various wavelengths can be used, but the most promising are UV and NIR resonance Raman spectroscopy. UV has low penetration in tissue giving the ability to examine superficial tissues while NIR penetrates more deeply.

Laser Doppler Imaging

LDI investigates laser light scattered by moving blood and static tissue. It operates on the same principle as ultrasonic Doppler, but involves very small tissue volumes and superficial sites only.

The laser Doppler imager scans a low power laser beam in a raster pattern over the skin or other tissue. Moving blood causes a Doppler shift which is processed to build up a colour coded image of blood flow (27). Unlike ultrasound, no skin contact is necessary and this allows the technique to be used to assess conditions which cannot be imaged with ultrasonic Doppler. It is therefore valuable for the assessment of the depth of major burns prior to plastic surgery and skin grafting, and to study the skin plaques of psoriasis (28). Perfusion of the extremities and joints in rheumatology and such conditions as Raynaud's disease are investigated with LDI. It is useful in the prediction of areas which may ulcerate in diabetic patients.

The technique can only be used in close proximity to the area to be studied, but the addition of flexible fibre optic probes opens the way to the study of internal structures.

New Uses for Lasers

New diagnostic and therapeutic applications of medical lasers are announced frequently. For instance an air-cooled argon laser has recently been used in dentistry to find areas of demineralisation in tooth enamel before a visible cavity appears. The green light is shone onto the tooth via an optical fibre and fluorescence from the surface is collected. Healthy enamel consists of a regular array of long thin hydroxyl apatite crystals. Light is guided through the array with little scatter into the dentine below. However, demineralised enamel contains many microscopic cracks which scatter light strongly giving more absorption and greater fluorescence. Areas of damaged enamel can be treated with fluoride to prevent the formation of cavities (29).

There is still enormous scope for development in the field of medical lasers.

References

1. Farr, R. F. and Allisy–Roberts, P. J., *Physics for Medical Imaging*, London, W. B. Saunders, 1997, ISBN 0-7020-1770-1.

2. Wells, P. N. T. (ed.), *Ultrasonics in Clinical Diagnosis*, London, Churchill Livingstone, 1972, ISBN 0-443-00904-X.

3. Dussik, K. T., Dussik, F. and Wyt, L., *Wien. Med. Wochenschr.* **97**, 425, 1947.

4. Ludwig, J. D., *J. Acoust. Soc. America* **22**, 862, 1950.

5. Wild, J. J. and Reid, J. M., *Science* **115**, 226–230, 1952.

6. Satomura, S., *J. Acoust. Soc. America* **29**, 1181, 1957.

7. Leskell, L., *Acta Chir. Scand.* **110**, 301–315, 1956.

8. McDicken, W. N., *Diagnostic Ultrasonics: Principles and Use of Instruments*, Chichester, John Wiley and Sons, 1990.

9. Meire, H. B. and Farrant, P., *Basic Ultrasound*, Chichester, John Wiley and Sons, 1995.

10. Rumack, C. M., *Diagnostic Ultrasound*, New York, Mosby, 1997, ISBN 0-8151-8683-5.

11. Langton, C. M., Palmer, S. B. and Porter, R. W., *New Eng. Med.* **13(2)**, 89–91, 1984.

12. Richardson, R. E. (ed.), *Guidelines for the Routine Performance Checking of Ultrasound Equipment*, York, IPSM, 1998, ISBN 0-904181-55-3.

13. Bloch, F., Hansch, W. W. and Packard, M. E., *Phys. Rev.* **69**, 127–129, 1946.

14. Andrew, E. R., *Brit. Med. Bull.* **40**, 115–119, 1984.

15. Rinck P. A., *An Introduction to Magnetic Resonance Imaging*, Oxford, Blackwell Scientific Publications, 1993.

16. Sigal, R., *Magnetic Resonance Imaging*, Heidelberg, Springer-Verlag, 1988.

17. Westbrook, C. and Kaut, C., *MRI in Practice*, Oxford, Blackwell Scientific Publications, 1993.

18. National Radiological Protection Board, *Statement on Clinical Magnetic Resonance, Diagnostic Procedures*, N.R.P.B., 1991.

19. Runge, V. M., Stewart, R. G. *et al.*, *Radiology* **147**, 789–791, 1983.

20. Buckley, D. L., Mussurakis, S. and Horsman, A., *J. Computer Assisted Tomography* **22**, 47–51, 1998.

21. Smith, S. R., Martin, P. A. *et al.*, *Br. J. Cancer* **61**, 485–490, 1990.

22. Negendank, W. (review article), *NMR in Biomed.* **5**, 303–324, 1992.

23. McDonald, P. and Strange, J., *Physics World*, **11(7)**, 29–34, 1998.
24. Jöbsis, F. F., *Science* **198**, 1264–1267, 1977.
25. Barr, H., Dix, T. and Stone, N. (review article), *Lasers in Med. Sci.* **13**, 3–13, 1998.
26. Symanski, H. A., *Raman Spectroscopy, Theory and Practice*, New York, Plenum Press, 1967.
27. Essex, T. J. H. and Byrne, P. O., *J. Biomed. Eng.* **13**, 189–194, 1991.
28. Speight, E. L., Essex, T. J. H. and Farr, P. M., *Br. J. Dermatol.* **128**, 519–524, 1993.
29. Graydon, A., *Opto Laser Europe*, Issue 47, 37, 1998.

FURTHER READING

In addition to the references listed at the end of each chapter, the reader may find the following book list useful.

CHAPTER 1

Bleich, A. R., *The Story of X-rays*, New York, Dover, 1960.

Eisenberg, R. L., *Radiology, An Illustrated History*, St. Louis, Mosby Year Book, 1992, ISBN 0-8016-1526-7.

Glasser, O., *Wilhelm Conrad Röntgen and the Early History of Röntgen Rays*, London, John Bale & Sons, 1933.

Mould, R. F., *A History of X-rays and Radium*, London, IPC Business Press, 1980.

Mould, R. F., *A Century of X-rays and Radioactivity in Medicine*, Bristol, IOP Publishing, 1993,

Thomas, A. M. K., *The Invisible Light — 100 Years of Medical Radiology*, Oxford, Blackwell, 1995, ISBN 0-86542-627-9.

Weber, R. L., *Pioneers of Science*, Bristol, IOP Press, 1980, ISBN 0-85498-036-9.

CHAPTER 2

Curry, T. S., Dowdry, L. E. and Murray, R. C., *Christensen's Physics of Diagnostic Radiology*, New York, Lea & Febiger, (4th ed.) 1992.

Faulkner, K. (ed.), *Physics in Diagnostic Radiology*, York, IPSM, 1990, ISBN 0-904181-60-X.

Harrison, R. M. and Isherwood, I. (eds.), *Digital Radiology*, York, IPSM, 1984, ISBN 0-904181-27-8.

Mettler, F. A. and Upton, A. C., *Medical Effects of Ionising Radiation*, London, W. B. Saunders, 1995.

Ring, E. F. J., Evans, W. D. and Dixon, A. S., *Osteoporosis and Bone Mineral Measurement*, York, IPSM, 1989,

Romans, L. E., *Introduction to Computed Tomography*, London, Williams & Wilkins, 1995.

Sutton, D. (ed.), *Textbook of Radiology and Imaging*, London, Churchill Livingstone, 1998, ISBN 0-443-05368-5 (2 volumes).

Wall, B. F., Harrison, R. M. and Spiers, F. W., *Patient Dosimetry Techniques in Diagnostic Radiology*, York, IPSM, 1988, ISBN 0-904181-49-9.

CHAPTER 3

Fowler, J. F., *Nuclear Particles in Cancer Treatment*, Bristol, Adam Hilger, 1981, ISBN 0-85274-521-4.

Greening, J. R., *Fundamentals of Radiation Dosimetry*, Bristol, Adam Hilger, 1981, ISBN 0-85274-519-2.

McKinlay, A. F., *Thermoluminescence Dosimetry*, Bristol, Adam Hilger, 1981, ISBN 0-85274-520-6.

Mould, R. F., *Radiotherapy Treatment Planning*, Bristol, Adam Hilger, 1981, ISBN 0-85274-504-4.

Starkschall, G. and Horton, J. (eds.), *Quality Assurance in Radiotherapy Physics*, Madison, Medical Physics Publishing, 1991.

Williams, J. R. and Thwaites, D. I. (eds.), *Radiotherapy Physics*, Oxford Medical Publications, 1993. ISBN 0-19-965316-9.

CHAPTER 4

Diffey, B. L., *Ultraviolet Radiation in Medicine*, Bristol, Adam Hilger, 1982.

Diffey, B. L., *Ultraviolet Radiation and its Medical Applications*, London, HPA, 1978, ISBN 0-904181-10-3.

Diffey, B. L. and Langley, F. C., *Evaluation of UV Radiation Hazards in Hospitals*, York, IPSM, 1986.

Hughes, D., *Hazards of Occupational Exposure to Ultraviolet Radiation*, London, Science Reviews, 1982, ISBN 0-905927-15-X.

Moseley H. and Haywood, J. K., *Medical Laser Safety*, London, IPSM, 1987, ISBN 0-904181-43-X.

Suess, M. J. and Benwell–Morrison, D. A., *Non-Ionising Radiation Protection*, WHO (Europe), (2nd eds.) 1989.

CHAPTER 5

Brucer, M., *A Chronology of Nuclear Medicine*, St.Louis, Heritage Publications, 1990.

Goldstone, K. E. (ed.), *Radiation Protection in Nuclear Medicine and Pathology*, York, IPSM, 1991.

Langmead, N. A., *Radiation Protection of the Patient in Nuclear Medicine*, Oxford Medical Publications, 1983.

Mettler, F. A., *Essentials of Nuclear Medicine Imaging*, London, W. B. Saunders, 1998, ISBN 0-7216-5121-6.

Moores, B. M., Parker, R. P. and Pullan, B. R., *Physical Aspects of Medical Imaging*, Chichester, Wiley, 1981.

Sharp, P. F., Dendy, P. P. and Keyes, W. I., *Radionuclide Imaging Techniques*, New York, Academic Press, 1985.

Spencer, R. P., *New Procedures in Nuclear Medicine*, CRC Press, 1989.

CHAPTER 6

Docker, M. F. and Duck, F. A. (eds.), *The Safe Use of Diagnostic Ultrasound*, London, BIR, 1991.

Fish, P., *Diagnostic Medical Ultrasound*, Chichester, John Wiley, 1990, ISBN 0-471-92651-5.

Hill, C. R., *Physical Principles of Medical Ultrasonics*, Ellis Horwood, 1986.

Lerski, R. A. (ed.), *Physical Principles and Clinical Applications of Nuclear Magnetic Resonance*, London, HPA, 1985, ISBN 0-904181-38-3.

Zweibel, W. J., *Introduction to Ultrasound*, London, W. B., Saunders, 1997, ISBN 0-7216-6947-6.

The areas of Medical Physics not covered in this volume include computer science, physiological measurement and medical engineering. It is difficult to recommend a suitable "starting point" volume for computing, because the subject is changing so rapidly. An excellent introductory volume to the other areas is

Brown, B. H. and Smallwood, R. H., *Medical Physics and Physiological Measurement*, Oxford, Blackwell Scientific Publications, 1981.

See also
Webster, J. G. (ed.), *Medical Instrumentation*, New York, Houghton Mifflin, 1978.

SUBJECT INDEX

NAME INDEX